Enrichment of Low-Grade Phosphorites with Organic Acids

Yerkebulan Raiymbekov

Enrichment of Low-Grade Phosphorites with Organic Acids

Selective Leaching and Sustainable Processing of Low-grade Phosphorites

Springer

Yerkebulan Raiymbekov
M. Auezov South Kazakhstan University
Shymkent, Kazakhstan

ISBN 978-3-032-02909-6 ISBN 978-3-032-02910-2 (eBook)
https://doi.org/10.1007/978-3-032-02910-2

This Springer imprint is published by the registered company Springer Nature Switzerland AG
The registered company address is: Gewerbestrasse 11, 6330 Cham, Switzerland

Foreword

The global demand for phosphorus fertilizers continues to rise as agriculture seeks to meet the nutritional needs of a growing population. However, phosphate resources are not unlimited, and increasing attention is being directed toward the valorization of low-grade ores that were previously discarded as waste.

In this context, the present monograph by Dr. Yerkebulan Raiymbekov addresses a pressing scientific and industrial challenge: how to recover valuable phosphorus from low-grade phosphorites using sustainable methods. The research combines rigorous experimentation with environmental consciousness, offering a novel pathway through organic acid-based beneficiation.

With Kazakhstan holding significant reserves of phosphate rock, the methodologies and results presented in this book are particularly timely. They not only reflect scientific rigor but also resonate with the broader goals of sustainable development and circular economy. The monograph stands as a valuable contribution to the global body of knowledge in phosphate processing and should inspire future research and industrial implementation.

M. Auezov South Kazakhstan University
Shymkent, Kazakhstan

Yerkebulan Raiymbekov

Preface

This monograph presents a systematic investigation into the enrichment of low-grade phosphorites using organic acids—a topic of growing importance in the context of sustainable resource management and green chemistry. The motivation for this work arose from the urgent need to utilize vast off-balance reserves of phosphorite in Central Asia, particularly in Kazakhstan, where traditional beneficiation methods have proven inefficient or economically unfeasible.

The book is structured to guide readers through the theoretical foundations of selective leaching processes, followed by detailed experimental studies using acetic, citric, and formic acids. Each chapter integrates thermodynamic modeling, chemical kinetics, and phase transformation analysis to present a complete understanding of how these organic agents can optimize phosphorus recovery. Case studies from three major phosphorite deposits—Zhanatas, Akzhar, and Shilisai—demonstrate the practical relevance of the approach.

This work is intended for researchers, graduate students, and industry specialists in the fields of mineral processing, chemical engineering, and environmental science. It is hoped that the presented data and insights will contribute meaningfully to the development of economically viable and environmentally friendly technologies for phosphate beneficiation.

Shymkent, Kazakhstan Yerkebulan Raiymbekov

Acknowledgements

The author expresses sincere gratitude to the Ministry of Science and Higher Education of the Republic of Kazakhstan for supporting this research within the framework of targeted funding programs (Zhas Galym project, AP15473115). Special thanks are extended to the staff and colleagues at M. Auezov South Kazakhstan University, whose technical and analytical support was essential throughout this study.

The author is also grateful to international collaborators, particularly researchers at the Technical University of Munich, for their input on spectroscopic analyses and phase characterization. Constructive feedback from peer reviewers and the editorial team at Springer is deeply appreciated.

Lastly, the author wishes to thank his family for their unwavering encouragement, patience, and belief in the value of this scientific endeavor.

Contents

Chapter 1
Introduction

Global demand for phosphorus is escalating in tandem with the need to secure food, energy, and critical materials for a growing population. More than 90% of the element is consumed as phosphate fertilizers, and it is now widely acknowledged that sustainable access to P-bearing ores is as strategically important as access to oil or rare earths. Kazakhstan's Karatau Phosphorite Basin ranks among the world's largest deposits, yet only a fraction of its geological endowment can be fed directly to wet-process phosphoric-acid plants or electrothermic furnaces. Open-pit operations leave behind tens of millions of tons of low-grade waste rock (≈11 to 13 wt% P_2O_5) whose carbonate-rich, silica-cemented texture frustrates conventional flotation or thermal decarbonization. Stockpiles continue to rise, occupying vast areas of arid steppe and immobilizing a potentially valuable national resource.

Previous beneficiation attempts—high-temperature calcination, tribo-electrostatic separation, direct acidulation, or dolomite flotation—have succeeded only at bench scale or have proven uneconomic once energy consumption, reagent loss, and gypsum disposal are internalized. Against this backdrop, selective leaching with weak organic acids has re-emerged as a conceptually attractive route: carboxylic ligands can dissolve Ca- and Mg-carbonates at ambient pressure while leaving the apatite lattice and quartz framework largely intact, thereby upgrading P_2O_5 in a single hydrometallurgical step and generating acetate liquor that is benign and potentially marketable.

The present monograph delivers the first systematic, deposit-specific evaluation of organic-acid beneficiation applied to three representative off-balance ores from the Karatau province: phosphate–silica layer rocks from Zhanatas, low-grade phosphorites from Akzhar, and low-grade phosphorites from Shilisai.

This monograph introduces a novel approach to the beneficiation of low-grade phosphorites by utilizing weak organic acids, such as acetic, citric, and formic acids, to selectively leach out carbonate impurities while preserving the phosphate mineral matrix. Unlike conventional methods that require high-temperature

Y. Raiymbekov, *Enrichment of Low-Grade Phosphorites with Organic Acids*,
https://doi.org/10.1007/978-3-032-02910-2_1

treatment or intensive flotation processes, the organic acid treatment operates under milder conditions and is based on the thermodynamically favorable interaction of carboxylic acids with carbonate minerals. This allows for a more environmentally friendly and targeted upgrade of phosphate ores that were previously considered unfit for industrial use.

A key contribution of this work lies in the comprehensive kinetic characterization of the leaching processes. Through the application of the Jander–Zeldovich model and Arrhenius analysis, the study demonstrates that the enrichment reactions are controlled by a diffusion mechanism, with activation energies in a range that confirms the efficiency of low-temperature leaching. The dissolution behavior of different mineral components is thoroughly examined, highlighting the selective removal of calcite and dolomite phases while retaining the structural integrity of apatite.

Furthermore, the work provides a detailed physicochemical profile of both the solid residues and the liquid by-products resulting from the leaching process. Advanced characterization techniques such as SEM-EDX, XRD, and FTIR reveal the preservation of the crystalline phosphate structure and offer insight into the residual mineralogy after leaching. Special attention is given to the composition of the acetate-rich filtrates, which contain calcium, magnesium, and aluminum acetates—compounds with potential for secondary valorization in the food, textile, and agricultural industries. This not only reduces the environmental burden associated with waste disposal but also opens avenues for circular economy integration in phosphate processing.

In essence, the study establishes the scientific basis for a selective, efficient, and potentially scalable method of enriching low-grade phosphorites. By combining thermodynamic, structural, and analytical insights, it advances both the theoretical understanding and the practical application of organic acid-based beneficiation techniques.

The practical importance of this research lies in its direct applicability to solving one of the most pressing challenges in phosphate resource management: the industrial utilization of low-grade phosphorites. Large quantities of such ores—particularly those from the Zhanatas, Akzhar, and Shilisai deposits—have long remained unused due to their insufficient P_2O_5 content and the inefficiency of traditional beneficiation technologies [1]. The proposed method of selective acid leaching using weak organic acids presents a viable alternative that can be implemented using relatively simple equipment and moderate operational conditions.

By demonstrating the ability of acetic, citric, and formic acids to selectively dissolve carbonate components, the research outlines a route to significantly increase the phosphorus content in low-grade ores, bringing them up to the standards defined in national technical specifications for thermal (FKT) and extractive (FKE) phosphate applications. For example, phosphorite concentrates from Shilisai treated with acetic acid were shown to reach P_2O_5 levels suitable not only for thermal phosphorus production, but also for phosphoric acid extraction. This dual applicability enhances their market value and allows for greater flexibility in downstream processing.

In addition, the study pays particular attention to the side products of acid treatment—namely, acetate salts formed in the liquid phase—which represent a secondary resource of economic interest. Through a combination of elemental analysis and phase identification, it was shown that these salts predominantly consist of calcium and magnesium acetates with favorable physicochemical properties. Such compounds have proven utility in food preservation, agriculture, and pharmaceuticals, suggesting that they can be repurposed following appropriate purification, thereby contributing to waste minimization and value-added recovery.

Importantly, the methods developed and validated in this work are not confined to laboratory settings; the reagents used are inexpensive, and the reaction conditions are compatible with industrial scale-up. The approach is particularly relevant for phosphate-producing countries with significant off-balance ore reserves and limited access to complex processing infrastructure. By integrating organic acid-based beneficiation into existing production frameworks, these countries can reduce their reliance on imported high-grade phosphates and improve the sustainability and profitability of domestic production.

In conclusion, the practical outcomes of this research offer new perspectives for phosphate beneficiation, waste reduction, and the development of multiproduct processing schemes. This contributes to national resource efficiency and aligns with global sustainable development goals targeting responsible consumption and production.

Chapter 2
Literature Review

2.1 Characteristics of Phosphorites and Their Resource Indicators

Phosphorus is one of the most abundant elements on Earth's surface and plays a crucial role in both geological and biological processes. In the field of mineralogy, phosphorites exhibit structural complexity and considerable diversity. The phosphorus element is an integral component of all biological and physiological processes in both plant and animal organisms. Additionally, phosphate minerals represent a valuable natural resource widely utilized as agricultural fertilizers [2].

Phosphorites occur in various geological environments and exhibit highly diverse mineralogical structures. Within igneous and metamorphic rocks, members of the apatite group, particularly fluorapatite, predominate as the primary phosphorus-bearing minerals [3]. Based on their formation process, phosphorites are classified into two genetic groups: marine and continental. Marine phosphorites, in turn, are further categorized according to lithological characteristics into fine-grained, granular, nodular, and shell-based types. These classifications are not arbitrary, as they reflect distinct geological formation features, stratigraphic positioning, and geographical distribution [4]. The primary classification of phosphorites is presented in Fig. 2.1.

Granular phosphorites are the most abundant among phosphorite classifications, accounting for approximately 70% of total reserves. Fine-grained phosphorites constitute around 12% and are predominantly found across the continents of North America, Asia, and Australia. Nodular and shell-based phosphorites collectively represent only 1–2% of the world's total phosphorite reserves.

According to the International Fertilizer Development Center (IFDC), a global map of phosphorite deposits has been developed based on their classification, as presented in Fig. 2.2.

Y. Raiymbekov, *Enrichment of Low-Grade Phosphorites with Organic Acids*,
https://doi.org/10.1007/978-3-032-02910-2_2

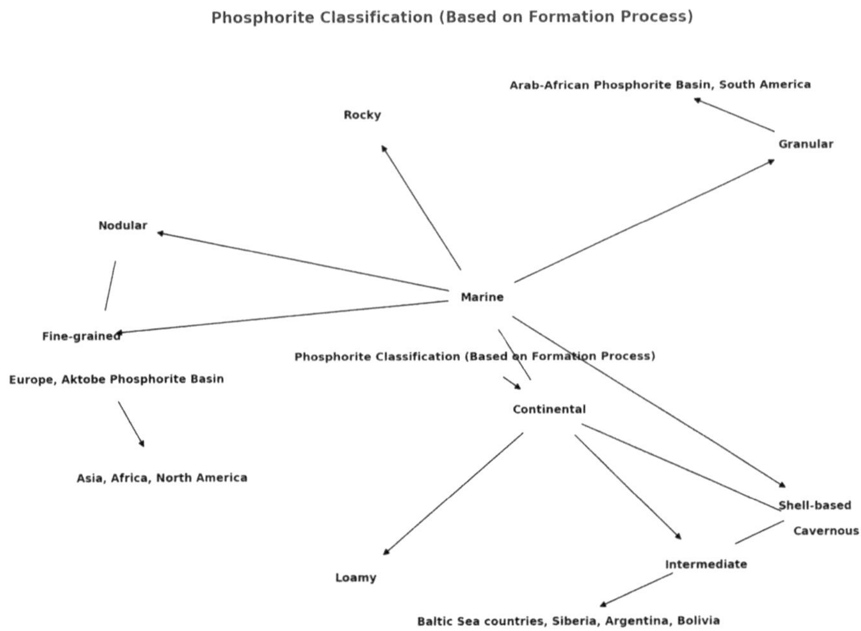

Fig. 2.1 Phosphorite classification

Fig. 2.2 Global distribution of phosphorite deposits by classification [5]

Data from the map indicate that sedimentary (marine) phosphorites are significantly more widespread than igneous and island-type phosphorite deposits. The primary countries holding the largest sedimentary phosphorite reserves include the United States, Morocco, China, and Russia. In Africa, particularly in Morocco, phosphorite deposits contain up to 35% P_2O_5, eliminating the need for additional beneficiation processes. In the United States, major phosphorite deposits are located in the Florida and North Carolina phosphorite basins, whereas in China, the majority of phosphorite reserves are concentrated in the Hubei-Hunan basin. The P_2O_5 content in these deposits generally ranges between 18% and 29% [6]. In neighboring Russia, the apatite deposits belonging to the Khibiny group are classified as some of the largest phosphorite reserves [7].

According to the U.S. Geological Survey (USGS) report from 2020 [8], a comprehensive list of the world's largest phosphorite deposits has been compiled, and studies have been conducted to assess their reserves. The results of this research are presented in Table 2.1.

The majority of phosphorite deposits in Africa are concentrated in Morocco. Additionally, several phosphorite deposits are found in the countries of the Western Sahara region, including Algeria, Tunisia, and Egypt. The total combined reserves of phosphorite deposits in these countries are estimated to be approximately 50,000 million tons, and they are considered economically viable due to their high quality [9].

One of the major phosphorite deposits in China, Guizhou, is located in the southern part of the country and contains reserves of approximately 801 million tons. This deposit is twice the size of the previously considered large Kaiyang deposit and covers an area of 50 square kilometers [10, 15].

In Australia, the Georgina, Queensland, and Northern Territory phosphorite deposits account for 90% of the total phosphorite reserves on the continent. However, Australian phosphorites are characterized by an excessive presence of aluminum and iron oxides, which significantly reduce their commercial value. Additionally, mechanical and flotation beneficiation methods are considered

Table 2.1 Global phosphorite deposits and their reserves (U.S. Geological Survey, 2020) [8, 9–14]

Country	Reserves (million tons)	Major deposits
Morocco (including Africa)	50,000	Khouribga, Bou Craa, Youssoufia, Abu Tartur, Mhamid
China	3200	Guizhou, Kaiyang, Kunming, Yunnan
United States	1000	Florida, North Carolina
Australia	1100	Georgina, Queensland, Clinton, Catford, Mokulta
Russia	600	Khibiny, Seligdar, Vyatka-Kamsky, Kirovsky
Saudi Arabia	1400	As-Sanam, Khuraymiz, Al-Amoud, Turaif-Sirhan
Kazakhstan	260	Akzhar, Geres, Koksu, Zhanatas, Shilisai

inefficient for processing these raw materials. As a result, the production of mineral fertilizers without chemical treatment has been widely adopted [11].

Neighboring Russia holds approximately 2–2.5% of the world's total phosphorite reserves. The country ranks fourth in terms of reserves and fifth in phosphorus production. The Khibiny phosphorite region contains 40% of Russia's total phosphorite reserves and includes nearly ten major nepheline-apatite deposits [12].

Due to its geographical location, Saudi Arabia belongs to the East-Mediterranean phosphorite basin, where large phosphorite deposits are predominantly found in the Turaif-Sirhan region. The first deposits were discovered in the 1950s–1960s and have been actively exploited ever since. These deposits are classified as granular phosphorites, embedded within interbedded layers of siliceous, clayey, and carbonate rocks [13].

The Karatau Phosphorite Basin serves as the primary raw material source for phosphorus compound production in Kazakhstan. On a global scale, it is classified among the world's largest phosphorite deposits. The Karatau Phosphorite Basin is located in the northeastern part of the Kishi Karatau Ridge in Southern Kazakhstan, extending northwestward for approximately 120 km with a width of 25 km.

The phosphorite reserves in the Karatau Basin amount to approximately 700 million tons of P_2O_5 [16]. The Karatau Phosphorite Basin, which holds a significant share of the world's phosphate reserves, serves as a leading exporter of phosphates and their derivatives. The majority of phosphorite reserves in this region consist of sedimentary rocks with a substantial presence of carbonate minerals. Based on their chemical composition and mineral-petrographic characteristics, these deposits are classified into five types (Table 2.2).

According to the data presented in Table 2.2, the Karatau phosphorites predominantly consist of unmetamorphosed ores, based on their mineralogical and structural properties. Notably, the Aksai and Zhanatas deposits contain phosphorites that fall into all three categories in terms of useful component concentration.

According to Bushinsky's research [18], the main mass of Karatau phosphorites is confined within terrigenous-siliceous-carbonate formations. Within this group, two types of phosphorite formations can be distinguished: shale-siliceous-dolomitic and quartzite-shale-limestone (where quartzites are understood as quartz sandstones). Most phosphorite deposits in Karatau belong to shale-siliceous-dolomitic formations. Additionally, phosphorite layers are underlain by clayey and siliceous-clayey shales and overlain by Tamdy dolomites, indicating the presence of rich phosphorite deposits within the siliceous-dolomitic formations. These deposits have formed under conditions of medium- and small-scale deposits as well as phosphate occurrences.

According to the Geology Committee of the Ministry of Ecology, Geology, and Natural Resources of the Republic of Kazakhstan [19], phosphorites in Kazakhstan are mainly distributed across five regions:

1. Karatau Phosphorite Basin (Turkistan and Zhambyl regions)
2. Aktobe Phosphorite Basin (Aktobe region)
3. Deposits in North Kazakhstan and Akmola regions
4. Deposits in the Aral Sea region (Kyzylorda region)
5. Deposits in Central Kazakhstan

Table 2.2 Classification of Karatau phosphorites based on structure [17]

Group	Type	Grade classification by useful component content	P_2O_5, %	Deposits
Unmetamorphosed ores	Monophosphate	Rich	28–32	Zhanatas, Kokzhon, Koksu
	Carbonate		23–27	Kokzhon, Geres, Himmelfarb
	Silicon-carbonate		23–24	Aksai (southeastern part), Zhanatas (southeastern part)
	Carbonate-silicon	Ordinary	20–23	Aksai (northwestern part), Tuyesay, Sholaktau, Koksu, Zhanatas
	Pelitomorphic carbonate-silicon		20–23	Zhanatas, Koksu, Akzhar, Ushbas
	Clay-silicon-laminated rock	Poor	17–19	Zhanatas, Koksu, Akzhar, Ushbas
	Silicon-laminated rock		17–19	Sholaktau, Aksai
Metamorphosed ores	Apatite	Rich	30–36	Tesiktas
	Carbonate-silicon	Ordinary	21–26	Sholaktau (western part)
	Silicon	Poor	17–19	Sholaktau (western part)

Based on data provided by the Geology Committee [19], the names, locations, and geological characteristics of the phosphorite deposits are as follows:

Akzhar Deposit—Located in the Karatau Range, extending 6 km from southeast to northwest. The phosphorite layer has a complex structure, consisting of phosphorites interbedded with sericite-siliceous shales. The average P_2O_5 content ranges from 14% to 19%, increasing to 21% towards the southeast.

Aksai Deposit—Situated 40 km northwest of Karatau city. The phosphorite-bearing formation comprises four sections: the northwestern part contains high-quality and siliceous phosphorites, with alternating layers of siliceous phosphorites and siliceous-carbonate rocks, where the average P_2O_5 content is 20–21%. The deposit is classified into two types: carbonate (southeastern) and siliceous-carbonate (northwestern).

Himmelfarb Deposit—Located 3 km southwest of the Zhanatas exploration center. The phosphorite layers become more complex towards their base and are composed of phosphate-siliceous rocks, with the P_2O_5 content increasing up to 25% towards the southeast.

Zhanatas Deposit—Contains 35% of the total identified reserves of the Karatau Phosphorite Basin. Its mineralogical-petrographic structure has been studied to a depth of 700 m, and the total deposit area is 21 km^2. Geologically and productively, it consists of four major sections: Burkitti, Toguzbai, Tsentralny, and Shoshkabulak. The average P_2O_5 content across all sections is 25–30%.

Zhetimshoky Deposit—One of the southernmost deposits of the Karatau Phosphorite Basin, extending 500 m. It consists of a surface phosphorite-siliceous layer and a lower phosphorite layer, with an average P_2O_5 content of 6–11% and 18–23%, respectively.

Arbatas Deposit—Located 4 km southwest of the Sholaktau Phosphorite Deposit. The phosphorites are mainly coarse-grained carbonate types, with frequent interbedding of phosphate-siliceous and phosphate-carbonate rocks. The average P_2O_5 content is 18–19%.

Geres Deposit—Extending 9 km in length and 600 m in width. The phosphorites occur as fine and medium-grained varieties, with phosphorus anhydride content averaging 23–26%.

Zhylan Deposit—Located in the central part of the phosphorite basin, extending 5 km. The phosphorite structure is heterogeneous, with fine- and medium-grained phosphorites in some areas and coarse-grained phosphorites in others. The average P_2O_5 content ranges from 19% to 25%.

Koksu Deposit—Constitutes 13% of the phosphorite basin's reserves, extending 13 km. The average P_2O_5 content in surface layers is 26–28%, gradually decreasing to 23–27% at greater depths.

Kotyrbulak Deposit—Located 5 km southeast of the Sholaktau Deposit. The phosphorite layers are weakly metamorphosed, with medium- to coarse-grained textures. The average P_2O_5 content is 17–26%.

Kokzhon Deposit—Located in the northwestern part of the Kishi Karatau Range, extending 40 km from southeast to north. Most of the structures consist of phosphatized siliceous-carbonate shales. The deposit is divided into four main sections: Araltobe, Kesiktas, Atkum, and Kystas. The chemical composition is variable: in the northwestern part, the average P_2O_5 content is 26–29%, whereas in the southeastern part, it decreases to 23–26%, with a corresponding increase in silicon content.

Sholaktau Deposit—Located 5 km south of Karatau city in Zhambyl Region. The mineralogical-petrographic structure consists of three main sections: phosphate-carbonate (P_2O_5 19–33%), phosphate-siliceous (P_2O_5 20–22%), upper surface phosphate layer (P_2O_5 8–13%). The surface layer is characterized by a high content of quartz, silica, and feldspar formations.

Another notable group of phosphorite deposits in Kazakhstan consists of algal (nodular) phosphorites, located within the Aktobe Phosphorite Basin in the Aktobe region. These deposits are developed through open-pit mining and are classified as clustered algal-type phosphorites, characterized by their lithological structure of flat-lying and horizontally oriented layers ranging from 0.3 to 1.5 m in thickness.

Based on the average concentration of phosphorus pentoxide (P_2O_5), these deposits are categorized as low-grade phosphorites [20, 21].

A distinguishing feature of algal phosphorites is the high degree of isomorphic substitution of the phosphate group with the carbonate (CO_3) group in their structure. This substitution leads to the decomposition of phosphate minerals by carbonic acid present in the carbonate group, rendering the phosphates significantly more soluble and reactive [21]. This characteristic makes algal phosphorites highly suitable for beneficiation and presents them as promising raw materials for the production of phosphate flour.

According to data from the Geology Committee of the Republic of Kazakhstan [19], the following are selected descriptions of key algal phosphorite deposits in the Aktobe region:

Shilisai Deposit—Located approximately 90 km from Aktobe city, covering a total area of around 1550 square kilometers. The phosphorites at this site exhibit nodular morphologies, formed by a sequence of phosphatic concretions, algal structures, and phosphate-bearing plates, with homogeneous yet variable morphologies. The average P_2O_5 content ranges from 16% to 18%.

Pokrovskoe Deposit—Covers an area of 657 square kilometers, situated to the west of the Shilisai deposit. Its phosphorites are dominated by carbonate compounds, and the phosphorus pentoxide concentration is comparable to that of Shilisai phosphorites.

Koktobe Deposit—Recognized as one of the major deposits within the Aktobe Phosphorite Basin, with an area of approximately 100 square kilometers. The phosphatic strata are presented in the form of algal structures cemented with phosphate material. The phosphorites have oolitic and amorphous textures, with average P_2O_5 concentrations ranging from 14% to 18%.

Novoukrainskoe Deposit—Comprises algal phosphorites interbedded with glauconite-quartz compounds, forming a distinct phosphatic stratum. The total area is 45 square kilometers, with average P_2O_5 concentrations between 15% and 19%.

Alga Deposit—Located in the northwestern part of the Shilisai deposit, approximately 15 km west of the Novoukrainskoe site, with an explored area of 500 square kilometers. The average P_2O_5 content in the phosphorites ranges from 12% to 14%.

Several phosphorite deposits have also been explored in the Northern and Central regions of Kazakhstan. However, due to their chemical and mineralogical composition, these deposits are currently considered unsuitable for phosphorus-based production processes. For instance, the Kerbai, Koksegir, Koksor, and Tastykol deposits in Northern Kazakhstan contain uranium oxide concentrations ranging from 0.0001% to 0.5%, and iron and aluminum oxides in the range of 4.5–6%. Additionally, secondary hydrothermal mineralization has occurred in the form of uranium resins within these sites. Lithologically, the phosphatic layers are oolitic and fine-grained, with carbonatized textures, and the average P_2O_5 content in these deposits ranges from 12% to 25% [19].

A similarly high concentration of aluminum oxides is observed in the Nugyrbek and Sarisai deposits in Central Kazakhstan. These phosphorite sites are distinguished by their unique structural characteristics, notably oolitic carbonaceous-siliceous and clayey-siliceous compositions. In terms of mineralogy, beyond the phosphate component (15–20%), a significant portion of alumophosphate phases is also present, accounting for up to 30% of the material.

The Kazaly phosphorite deposit, located in the Kyzylorda region, comprises three primary areas: Tarsuak, Baikozha, and Bostobe. The phosphorites here are marked by a coarse-grained quartz-glauconitic structure with a characteristic greenish-brown coloration. Phosphatic material is predominantly concentrated in the lower conglomerate-bearing horizon and the upper algal-bearing layer. The total area of the deposit is approximately 50 square kilometers, with average P_2O_5 concentrations ranging from 13% to 15% [19].

Due to the limited geological and mineralogical investigation conducted on the phosphorite deposits of Northern and Central Kazakhstan and the Kyzylorda region, comprehensive and conclusive data about these reserves remain unavailable at present.

In addition to the aforementioned natural deposits, there has been a progressive increase in the accumulation of technogenic mineral formations (TMFs) in the Karatau Phosphorite Basin since the 1950s–1960s, as a result of open-pit mining operations. The excavation of phosphorites using surface mining techniques generates large volumes of overburden and surrounding barren rocks, which are typically deposited in external dumps. While the deep phosphatic strata are extracted, the upper layers, including topsoil and stripping rocks, are stored separately. These formations are predominantly composed of phosphate-siliceous-carbonate interbedded rocks.

A generalized process flow diagram for the processing of phosphorite ores from this region is presented in Fig. 2.3 [22].

Phosphate-siliceous interbedded rocks are considered non-balance ores and constitute the primary component of the technogenic reserves of Karatau phosphorites. These rocks are characterized by the predominance of siliceous and clay matrices over the total phosphatic mass (oolites and phosphate grains). Substantial accumulations of these phosphate-siliceous interbedded rocks are observed at designated dumps in the Zhanatas and Koksu deposits (*see* Fig. 2.4). For example, in the "Central" dump of the Zhanatas deposit alone, the volume of phosphate-siliceous interbedded rocks exceeds 21 million tons.

According to the research findings cited by Kiperman [17], information regarding the chemical composition and total volumes of phosphate-siliceous-carbonate interbedded rocks is provided in Tables 2.3 and 2.4.

Based on their chemical and mineralogical characteristics, these technogenic reserves are classified into three main categories: clay-siliceous, phosphate-siliceous, and phosphate-siliceous-carbonate interbedded rocks. The phosphate-siliceous interbedded rocks are characterized by the dominance of siliceous and clay matrices over the phosphatic components (e.g., oolites and phosphate grains).

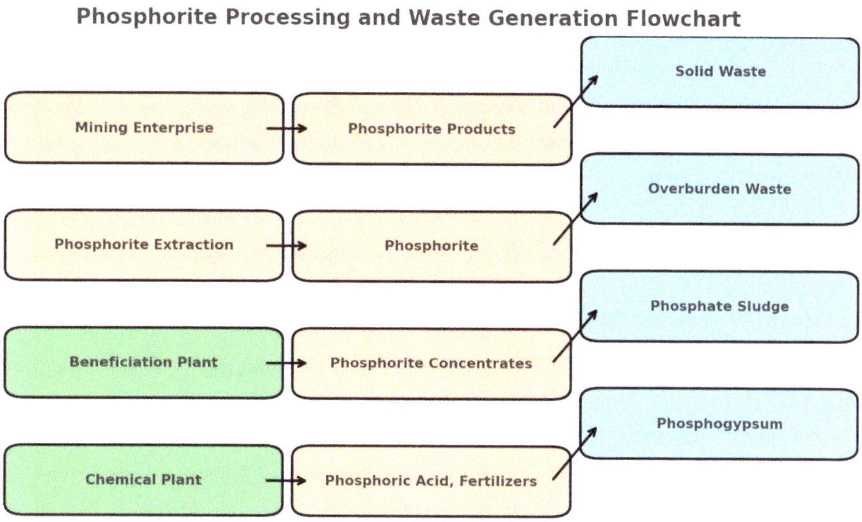

Fig. 2.3 Generalized process flow diagram for phosphorite rock processing: products and associated solid waste [22]

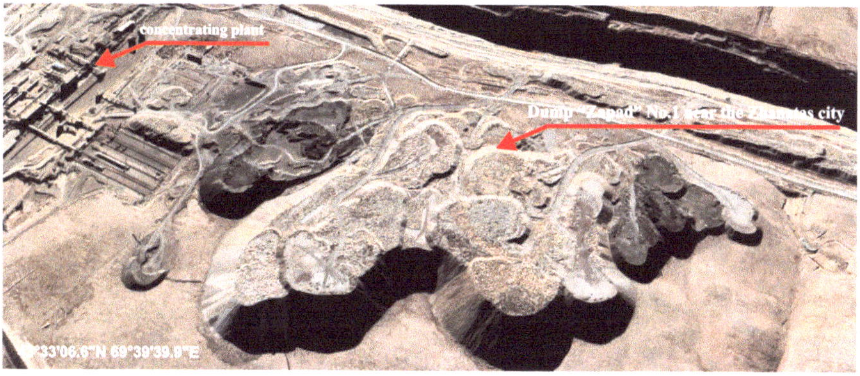

Fig. 2.4 "Zapad" No. 1 waste dump in the Zhanatas deposit area. (Based on Yandex map)

In contrast, the phosphate-siliceous-carbonate interbedded rocks contain a significantly higher proportion of carbonate compounds.

Due to these mineralogical and compositional features, the direct industrial application of such technogenic mineral formations is not feasible without prior beneficiation. This limitation is primarily attributed to the low content of the target component—phosphorus pentoxide (P_2O_5)—coupled with excessive concentrations of associated impurities, such as silicon dioxide (SiO_2), iron oxides (Fe_2O_3), and aluminum oxides (Al_2O_3).

Table 2.3 Characteristics of selected technogenic mineral formations in the Karatau Phosphorite Basin

Dump name	Volume (million tons)	P_2O_5, %	Insoluble residue, %	MgO, %
Zhanatas Deposit—"Zapad" Dump No. 1	2.4	11.90	55.00	2.40
Zhanatas Deposit—"Vostok" Dump No. 2	6.2	13.40	48.70	1.60
Zhanatas Deposit—"Zapad" Dump No. 2	2.1	14.20	50.20	1.80
Koksu Deposit—Special Dump No. 8	0.094	12.20	50.49	3.17
Koksu Deposit—Special Dump No. 10	15.3	8.00	58.69	1.54
Koksu Deposit—Special Dump No. 1	0.4	16–17	35–40	1.50–1.80
Kokzhon Deposit—Special Dump No. 6	0.886	11.94	52.40	2.40
Tuyesay Deposit—Special Dump No. 4	2.2	16–17	29–31	4.50–4.70

Based on the data presented above, it is evident that Kazakhstan possesses substantial phosphorite reserves. Currently, the major industrial exploitation of primary deposits within the Karatau Phosphorite Basin is being carried out by Kazphosphate LLP. At the Karatau Mining and Processing Complex, raw materials required for the production of yellow phosphorus and extraction-grade phosphoric acid are mined using open-pit methods from the Kokzhon, Koksu, and Zhanatas deposits, followed by mechanical beneficiation techniques [23].

However, the reserves of these actively exploited deposits are becoming increasingly depleted, and the exploration and development of new deposits require significant financial investment. Furthermore, the volume of phosphate-bearing waste generated during both mining and downstream processing continues to rise, while utilization strategies for these residues remain unimplemented. The use of low-grade phosphorite deposits, characterized by a low P_2O_5 content, is currently not viable for industrial application.

In the Aktobe Phosphorite Basin, the main Shilisai deposit is being processed by Temir-Service LLP, in accordance with GOST 5716-74 and ST LLP 930640000252-01-2011, to produce "CHILISAI"-branded phosphorite flour [24]. The primary constituent—phosphorus pentoxide (P_2O_5)—ranges between 17% and 18%, indicating that this deposit also falls within the category of low-grade phosphorites.

In this context, there is an urgent need to develop effective beneficiation technologies for low-grade phosphorites [25, 26]. The subsequent section of this literature review provides a comprehensive analysis of global and domestic research efforts focused on the beneficiation of phosphorite raw materials, including various methods developed and tested across different periods.

Table 2.4 Resource estimates of phosphorite deposits in Kazakhstan and associated technogenic reserves formed during open-pit mining

Name of phosphorite deposit	$A + B + C_1$ (P_2O_5-based), million tons	C_2 (P_2O_5-based), million tons
Karatau Phosphorite Basin		
Akzhar	5.07	16.8
Aksai	3.31	7.5
Himmelfarb	86.4	64.3
Zhanatas	133.2	–
Geres	7.2	0.7
Zhylan	–	48.0
Koksu	39.8	40.9
Kokzhon	13.08	2.3
Sholaktau	4.8	1.2
Technogenic mineral formations[a]	40.0	–
Aktobe Phosphorite Basin		
Shilisai	90.5	43.6
Pokrovskoe	202.3	108.6
Koktobe	25.6	–
Novoukrainskoe	6.6	–
Alga	6.2	93.7
Northern Kazakhstan[b]		
Tastykol	10.0	–
Central Kazakhstan[b]		
Nugyrbek	–	12.0
Sarisai	–	4.6
Kyzylorda Region[b]		
Kazaly	–	126.0

[a] Technogenic mineral formations refer to waste-derived reserves accumulated during open-pit operations

[b] Data for Northern, Central Kazakhstan, and Kyzylorda Region remain incomplete due to insufficient geological investigation

2.2 Methods of Enrichment of Low-Grade Phosphate Raw Materials

In global practice, several methods are used for phosphorite beneficiation, including flotation, separation in heavy suspensions, and mechanical processing techniques. Additionally, new beneficiation methods such as calcination, magnetic separation, and acid treatment are also being developed.

Mechanical beneficiation of phosphorites involves the use of various types of drums, sieves, and crushers with classifiers [27–30]. For instance, at the Karatau Mining and Processing Complex, phosphorites suitable for the needs of

Kazphosphate LLP undergo mechanical cleaning and crushing. During the Soviet era, research on the flotation beneficiation of phosphorites was conducted at this facility, with experimental trials carried out. For carbonate-type phosphorite raw materials, flotation was performed in an acidic medium, using phosphoric acid as a depressant reagent to suppress phosphate minerals. To separate carbonate compounds into the froth product, synthetic fatty acids were applied (Fig. 2.5) [17]. However, the application of carbonate flotation did not achieve the intended results. This was due to the fact that the grain size of phosphorite particles plays a crucial role in the beneficiation process, affecting both phosphorite and carbonate flotation.

In the beneficiation of low-grade phosphorite ores from the Karatau Phosphorite Basin, several flotation methods have been applied, including two-stage carbonate-phosphate flotation, suspension-flotation, and phosphate flotation. As flotation reagents, synthetic fatty acids, higher alcohols, liquid glass, kerosene, and sodium salts of aliphatic polyphosphoric acids (marketed as "Flotol-7.9") were used [31–34]. However, the low recovery rate of the target product, the formation of sludge waste, and the inability to recycle or dispose of the reagents demonstrated that flotation beneficiation of Karatau phosphorites is ineffective.

For instance, significant flotation beneficiation waste deposits have accumulated in the areas of Karatau and Sholaktau Mining and Processing Complexes since the 1960s–1970s. Specifically, the Zhanatas suspension beneficiation plant alone contains 350,000 tons of flotation waste with an average P_2O_5 content of approximately 15%. Additionally, the Karatau flotation beneficiation plant has 13 million tons of secondary waste, with P_2O_5 concentrations ranging from 14% to 17%.

To date, no effective solutions have been developed for the beneficiation and utilization of these secondary phosphate wastes in phosphate-related industries.

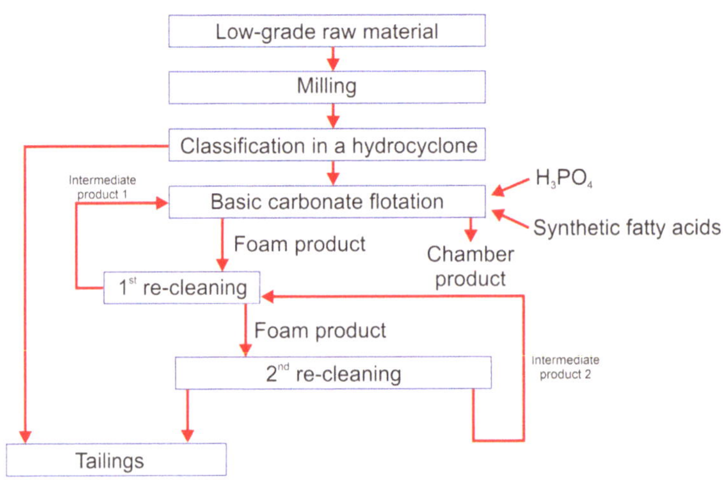

Fig. 2.5 Process flowchart for carbonate flotation implementation (Karatau Deposit, Kazakhstan) [17]

The carbonate-phosphate flotation method has been utilized in the beneficiation of phosphorites from the Minas Gerais deposit in Brazil [35]. In an acidic medium, phosphoric acid, citric acid, and carbon dioxide were used as depressants for phosphate minerals, while Flotanol D25, a flotation reagent based on polyalkylene glycol ether, was employed. Initially, the raw material contained approximately 17% phosphorus pentoxide (P_2O_5) and 20% calcium carbonate. However, after the flotation process, the P_2O_5 content increased to 32%, while the concentrations of calcium carbonate and quartz were significantly reduced.

Nevertheless, it is important to note that the flotation process also generates residual compounds in the form of sludges. Studies have revealed that these beneficiation residues still contain approximately 7% P_2O_5. Additionally, the high market cost of phosphorus and citric acid, along with the lack of consideration for their regeneration and reuse, highlight the limitations of the proposed technology. In this regard, phosphoric acid-based depressants have demonstrated their effectiveness as selective reagents in numerous studies [36–39]. However, in some cases, the amount of phosphoric acid required for the beneficiation of 1 ton of phosphorite reached 8–10 tons [40].

Another Brazilian deposit, the Itataia phosphorites, has also been studied in relation to flotation-based beneficiation [41]. In these studies, coconut oil and corn starch were used as flotation reagents. The results indicated that increasing the dosage of flotation reagents improved the separation efficiency of carbonate and silicate minerals. However, the characteristics of the generated waste, as well as the regeneration and disposal of the reagents, were not addressed in the study.

Plant oil-based flotation reagents have also been employed by Chinese researchers for the beneficiation of low-grade phosphorites from the Guizhou deposit [42]. In comparative trials, cottonseed oil, rice bran oil, rapeseed oil, and corn oil were used as anionic collectors during the flotation beneficiation process. Although the results demonstrated a notable increase in P_2O_5 content, the high concentration of dolomite minerals in the ore was identified as a limiting factor.

Researchers at the Lianyungang Chemical Engineering Design and Research Institute proposed the application of anionic collectors for the separation of dolomite minerals in phosphorites from the Yichang deposit [43–45]. Sodium dodecyl sulfate was employed as an anionic collector, leading to a P_2O_5 increase of up to 32%. However, the researchers concluded that the use of sodium dodecyl sulfate during flotation caused interactions with phosphate minerals and led to foam instability, which ultimately resulted in the abandonment of this method.

Studies utilizing anionic collectors [46–49] have demonstrated that the efficiency of compounds such as disodium dodecyl phosphate, sulfosuccinate, sulfosuccinamate, and hydroxamates is highly dependent on the solubility of mineral phases in phosphorites and their stability constants. For instance, the application of surface-active hydroxamate agents has proven effective in the beneficiation of Florida phosphate ores [50].

While long-chain fatty acids and their salts function as anionic collectors, amphoteric surfactants serve as cationic collectors in selectively targeting carbonate and silicate minerals associated with phosphate matrices. Notable examples of such reagents include *N*-sarcosine [51, 52], carboxymethyl imidazoline derivatives [53], and the sodium salt of aminopropionic acid [54, 55], among others.

The mechanism of cationic flotation of phosphorites using amine-based collectors operates on the principle that, under depressant action, the phosphate phase becomes suppressed, allowing positively charged amine species to interact electrostatically with negatively charged silicate surfaces, thereby achieving effective separation (Fig. 2.6).

Despite these advancements, the industrial application of both anionic and cationic collectors remains constrained due to a number of critical limitations. These include their high sensitivity to temperature fluctuations, substantial reagent consumption rates, elevated costs, and the generation of environmentally problematic residual waste [57]. Furthermore, the flotation of finely dispersed phosphate particles introduces additional challenges. The presence of calcium and magnesium carbonates, which exhibit flotation behaviors similar to phosphates, complicates the

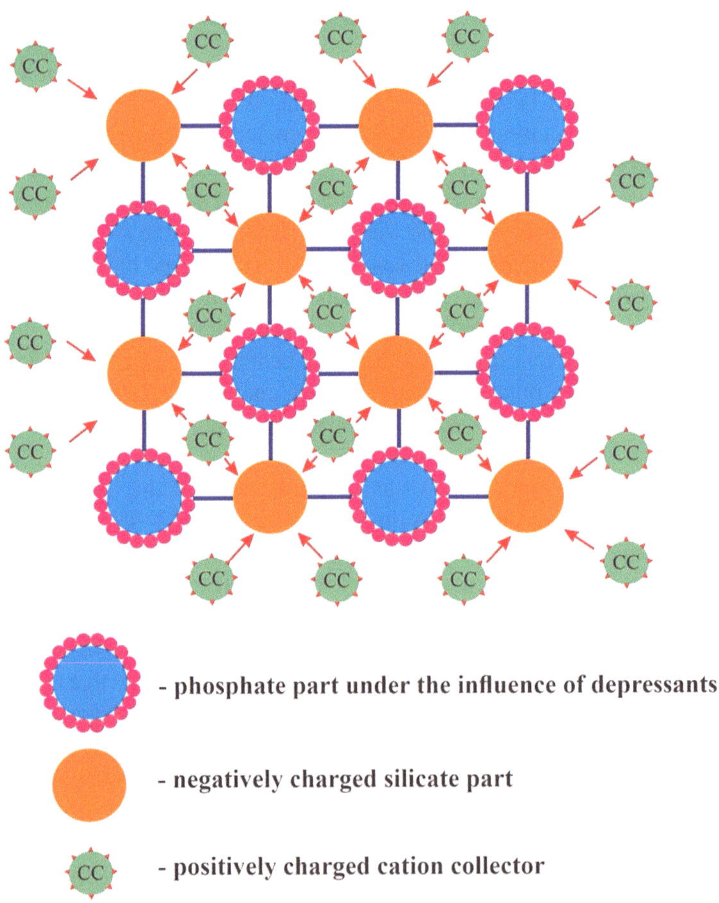

Fig. 2.6 Interaction mechanism of cationic collectors with silicate phase in phosphorite matrix [56]

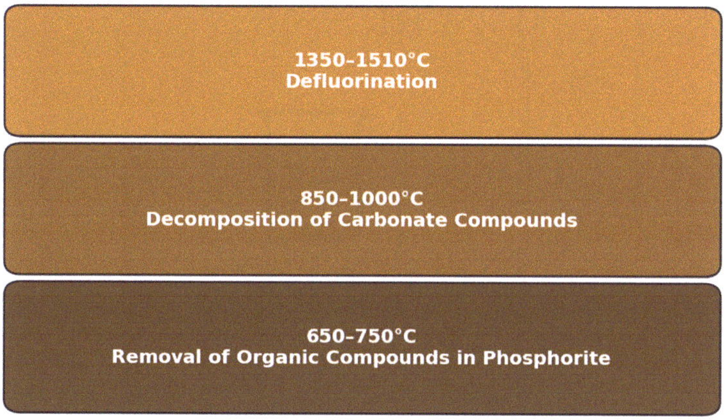

Fig. 2.7 Thermal transformations in phosphate feedstock under various calcination temperatures [59–62]

separation process. Aluminum and iron oxides further interfere with flotation performance, often resulting in contamination of the concentrate. As a consequence, the overall recovery yields are typically low, and the enrichment of valuable components in the final product remains unsatisfactory. These issues are compounded by the high cost of flotation reagents and the complexity of the associated technological and equipment requirements. Consequently, flotation-based beneficiation methods for phosphorites are frequently regarded as both economically and technologically inefficient.

The calcination method is widely utilized to eliminate carbonate and organic compounds present in phosphorite ores. Globally, over 10% of the phosphate raw materials are processed using this thermal treatment approach [58]. At temperatures between 650 and 750 °C, organic matter is decomposed through an exothermic reaction, primarily due to the combustion of organic residues and carbon content [59]. When the temperature is increased to the range of 850–1000 °C, carbonate minerals undergo thermal decomposition via an endothermic process [60, 61]. Further elevation of the calcination temperature to 1350–1510 °C initiates defluorination, during which fluoride compounds are removed and the crystalline lattice of phosphorus is disrupted, transforming phosphorus into a biologically available form [62]. As a result, defluorinated phosphates produced under these conditions are considered suitable for use as feed additives for livestock (see Fig. 2.7).

During the calcination process, calcium carbonate undergoes thermal decomposition to yield calcium oxide (CaO) and carbon dioxide (CO_2) (Fig. 2.8, Stage I). This transformation results in a notable increase in the $CaO:P_2O_5$ ratio within the phosphate material (Fig. 2.8, Stage II), which subsequently leads to an increased consumption of sulfuric acid during the acidulation of calcined phosphorite. As a consequence, the volume of phosphogypsum generated as a by-product also rises (Fig. 2.8, Stage III) [63].

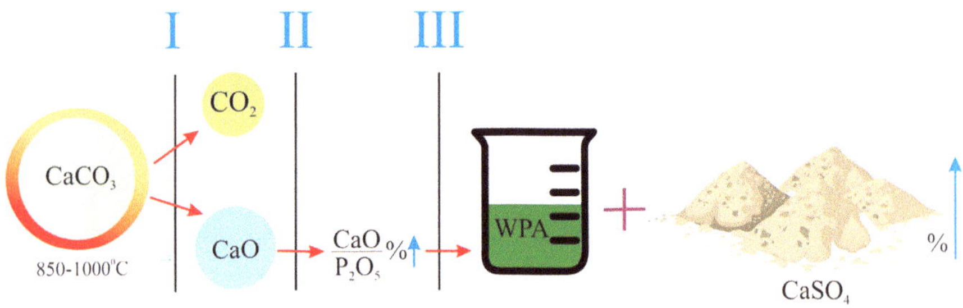

Fig. 2.8 Schematic representation of reactions during calcination of carbonate compounds in phosphate raw materials [63]

In response to these challenges, recent scientific and technological efforts have focused on developing strategies to mitigate the adverse effects associated with CaO accumulation. For instance, international researchers have proposed the use of ammonium chloride solutions to bind excess calcium oxide, resulting in the formation of calcium chloride ($CaCl_2$) as a stable residue, thereby reducing the surplus CaO content in the phosphate feedstock [64–66].

An alternative method for the removal of calcitic impurities from phosphate raw materials involves the wet slaking of calcined material, during which CO_2 gas is passed through the system until the transformation $CaO \rightarrow Ca(OH)_2 \rightarrow CaCO_3$ is completed. The resulting calcium carbonate suspension is then precipitated and separated from the treated material [67]. Additional methods proposed for the removal of calcium and magnesium carbonates released during calcination include the use of hydrocyclone separation with water [68–70], carbon dioxide gas [71], ammonium hydroxide [72], and even sugar-based solutions [73].

On a global scale, the calcination method gained industrial relevance during the 1960s and 1970s, when large-scale facilities were implemented. For instance, at the Jebel Onk deposit in Algeria, three calcination units were installed with an annual capacity exceeding one million tons of beneficiated phosphate concentrate. A similar approach is used by the Office Chérifien des Phosphates (OCP) in Morocco. In the United States, the method has also been widely adopted, with approximately 17 calcination units in states such as Idaho, Wyoming, and North Carolina, processing over 20 million tons per year of low-grade phosphates [74–80]. In Saudi Arabia, the Ma'aden Phosphate Company has conducted pilot studies on applying the calcination process to the Al-Jalamid deposit [81, 82]. However, these studies revealed that the operational costs, technical challenges associated with high carbonate content, inferior product quality, and high energy consumption render the method economically unfeasible.

Moreover, recent studies have identified additional limitations of the calcination process [83, 84]. Notably, high-temperature calcination reduces the solubility and reactivity of phosphates. In other words, calcined phosphorite requires extended post-treatment exposure in order to regain the solubility levels of untreated phosphorite. This phenomenon may be attributed to structural and chemical transformations occurring at elevated temperatures. For example, it has been observed that chlorine compounds in the phosphate matrix exhibit significantly reduced solubility after calcination [85]. Incomplete decomposition of carbonate and dolomitic components is also problematic, as their decomposition products can interfere with subsequent sulfuric acid digestion. Therefore, increasing the duration of calcination at 900–950 °C, or raising the calcination temperature threshold, is often necessary. However, this can also lead to over-calcination or melting of phosphate phases [86]. Additionally, the decomposition of dolomite at high temperatures may produce magnesium oxide, which subsequently transforms into insoluble magnesium hydroxide, further complicating the processing and utilization of calcined material in wet-process phosphoric acid (WPA) and fertilizer production systems [87].

Researchers have proposed the use of biotechnological approaches as an effective method for decomposing carbonate and dolomitic compounds found in low-grade phosphate ores. The use of microbial cultures from *Rhizopus*, *Aspergillus*, and *Candida* yeasts has demonstrated promising results in solubilizing calcium and magnesium carbonates during phosphate beneficiation. Application of these microorganisms has led to an increase in P_2O_5 content in phosphate concentrates from an initial 22% up to 31% [88]. Similar studies have recommended the use of *Thiobacillus*, *T. ferrooxidans*, and *T. thiooxidans* strains [89]. The metabolic activity of these microorganisms can be optimized by adjusting various technological parameters, including temperature, pH, and oxygen supply [90–92]. However, the main limitation of this approach lies in the complexity of microbial synthesis and its integration into industrial processes.

In addition to microbial methods, researchers have also explored the potential of acid leaching techniques for dissolving carbonates and associated impurities in phosphate ores. For instance, treatment of high-carbonate phosphorites with nitric acid has been shown to increase P_2O_5 content in the concentrate from 18% to 31% [93]. Another reported method involves the use of hydrochloric acid solutions (5–7%) at 20–60 °C to leach Karatau phosphorites with an initial P_2O_5 content of 20%, resulting in an increase to 26–27% [94]. A variant of this method includes a two-step process where low-grade phosphorite is first leached with hydrochloric acid under similar conditions, and the resulting concentrate is subsequently treated with phosphoric acid. An advantage of this technique is that the spent HCl solution can be regenerated using sulfuric acid, improving process sustainability. This combined approach has demonstrated a further increase in P_2O_5 content up to 28–34% [95].

To further enhance phosphate concentration, researchers have used 4–7% sulfuric acid solutions in three-stage acid leaching processes, which enable the dissolution of up to 99% of calcium and magnesium carbonates. As a result, P_2O_5 content in the concentrate can reach 25%, while the magnesium content can be reduced to as low as 1% [96]. Nevertheless, while mineral acids are effective chemical agents, their use presents certain drawbacks. These include partial loss of P_2O_5 into the leachate during processing [97–99], the high cost of acids, and limited strategies for reagent recovery and reuse, all of which challenge the economic viability of acid-based beneficiation techniques.

In addition to the use of mineral acids, recent studies have also explored the application of various inorganic salts in the beneficiation of phosphate ores. One such study demonstrated that a mixture of ammonium chloride and ammonium nitrate solutions, applied at 100–205 °C, could increase the P_2O_5 content in phosphate concentrates from 17.25% to 30%. The authors also proposed a method for reusing the spent leachate following the leaching process [100]. Furthermore, the use of solutions containing alkaline metal or ammonium-substituted phosphates has been reported to raise the P_2O_5 content to approximately 26% [101]. In another study, the application of 10–20% urea (carbamide) solutions led to an increase in P_2O_5 concentration in the phosphate concentrate to 36–44%. However, it was noted that in liquid-phase processing, calcium phosphate is co-extracted along with urea [102]. Despite their promising results, these methods are considered industrially nonviable due to insufficient validation, losses of the phosphate fraction, and the high cost of reagents.

Currently, one of the most promising emerging techniques for the treatment of low-grade phosphate ores is leaching with weak organic acid solutions. These acids exhibit high efficacy in carbonate dissolution, are cost-effective, technologically simple, and offer potential for reagent recycling, making them attractive alternatives to conventional methods. Literature review reveals the frequent use of formic, acetic, lactic, tartaric, citric, and maleic acids in experimental studies. A key advantage of these organic acids lies in their selectivity for dissolving carbonate compounds within phosphate ores.

In the former USSR, researchers proposed a two-step method for reducing the content of calcium and magnesium carbonates in phosphate raw materials. The method involved pre-calcination at 900 °C, followed by leaching with a 2% maleic acid solution at 50 °C for 20 min [103]. In this process, calcium and magnesium carbonates readily reacted with maleic acid to form soluble maleate salts, while the phosphate component remained largely in the solid phase. As a result, the P_2O_5 content of the phosphate concentrate increased from 26% to 28%, with only about 2.3% P_2O_5 lost to the leachate. The authors attributed this limited loss to physicochemical transformations in the phosphate matrix caused by pre-calcination.

Further investigations into the effectiveness of low-molecular-weight monobasic carboxylic acids for carbonate dissolution in phosphate ores demonstrated that P_2O_5 was fully retained in the solid phase, without any detectable transfer into the liquid phase [104].

Among monobasic carboxylic acids, formic acid (HCOOH) has garnered particular attention. One of its distinguishing features compared to other organic acids is its low molecular weight, which enhances its ability to rapidly penetrate the porous structure of phosphate ores, thereby accelerating reaction kinetics [105]. A study investigating the influence of formic acid concentration, leaching temperature, duration, and liquid-to-solid (L:S) ratio demonstrated an increase in P_2O_5 content from 26% to 34% under optimized conditions [106]. These optimal parameters included a leaching time of 25 min, an L:S ratio of 1:4, and a formic acid concentration of 4.5%. Additionally, the researchers observed that increasing the acid concentration beyond 8% led to phosphate loss into the leachate, and that extending the reaction time had little impact on the selective dissolution of carbonates.

A related study [105] confirmed that the optimal parameters for phosphate enrichment with formic acid are a concentration of 5–6%, leaching time of 25 min, and a temperature of 50 °C. Beyond this temperature threshold, the solubility and reactivity of formic acid decline, rendering it ineffective for dissolving carbonate impurities.

In addition to formic acid, other carboxylic acids, such as lactic acid and succinic (ethanedioic) acid, have been evaluated in the beneficiation of low-grade Pakistani phosphorites [107, 108]. At 45 °C, using an L:S ratio of 1:7 and an acid concentration of 8%, the resulting phosphate concentrate showed an increase in P_2O_5 content to 33–34%.

Among organic monocarboxylic acids, acetic acid has been one of the most frequently studied for carbonate dissolution in low-grade phosphorites. Its wide availability and low market cost are key advantages for industrial-scale application. Table 2.5 provide a general overview of the leading global techniques employed for the beneficiation of phosphate raw materials, including a summary of acetic acid-based leaching methods applied to various low-grade phosphate ores.

According to the data presented in Table 2.5, the optimal concentration of acetic acid varies depending on the characteristics of specific phosphate deposits, typically ranging from 3% to 6%. The duration and temperature of the leaching process are generally consistent across studies. However, the gap between the initial and final P_2O_5 contents depends on the mineralogical composition of each deposit. In particular, phosphorites with a high carbonate content tend to exhibit greater increases in P_2O_5 content after treatment, while those rich in silicate minerals demonstrate more modest gains.

Table 2.5 Global practices using acetic acid for low-grade phosphate ore beneficiation [109–120]

Phosphate deposit	Acetic acid concentration	Solid:liquid ratio	Process temperature and duration	P_2O_5 content (%) before → after
Ruseifa, Jordan	6–7%	1:5	25–30 °C, 30–35 min	24–25 → 30–32
Hazara, Pakistan	6%	1:5	45 min	24 → 30
Mardin-Mazidagi, Turkey	3%	1:5	22 °C, 40 min	12.7 → 29
Abu-Tartur, Egypt	0.8 M	1:4	20–25 °C, 50 min	21.8 → 30
Guizhou, China	5%	1:40	40 °C, 30 min	17 → 28–30
Akash, Iraq	2%	1:5	30 °C, 20 min	–
Khouribga, Morocco	7%	1:3	40 °C, 60 min	18 → 30
Epirus, Greece	1 M	1:2.5	25 °C, 60 min	13 → 31
Dzheroy-Syrdarya, Uzbekistan	4–6%	1:8	25 °C, 30 min	15 → 22

For instance, research conducted on the phosphorites of the Ruseifa deposit in Jordan [109, 110] identified that the mechanical hardness and fineness of phosphate rocks significantly affect the leaching efficiency. Moreover, complete dissolution of carbonate phases depends not only on the reaction time and temperature but also on the particle size of the phosphate feedstock. Nevertheless, the same studies concluded that extending the reaction duration beyond 30–35 min does not result in any further improvement in leaching efficiency.

In studies conducted by Pakistani researchers, it was suggested that phosphate ores should undergo a preliminary flotation stage to reduce the content of silicate compounds before acid treatment [111]. As in the previous studies, these researchers confirmed that extending the leaching duration or increasing the temperature beyond optimal values is unnecessary. Elevating the temperature leads to the release of carbon dioxide, which may contaminate the leaching medium by interacting with acetic acid and water vapor, ultimately reducing the solvent efficiency of the acid.

The Mardin-Mazidagi phosphate deposit in Turkey is classified as highly carbonated, with carbonate contents reaching up to 30%, and an average P_2O_5 concentration of around 12.7%. During experimental acid leaching, an increase in acetic acid concentration initially resulted in a corresponding rise in P_2O_5 content. However, beyond a certain threshold, further acid concentration increases led to a decrease in the P_2O_5 content. This phenomenon was attributed by the researchers to the diffusion of hydrated calcium ions into the bulk solution after approximately 40 min of reaction. They concluded that highly concentrated organic acids tend to exhibit more limited diffusion capabilities [112].

Despite the elevated concentrations of carbonate compounds, the Abu-Tartur phosphate deposit in Egypt demonstrated a ~9% increase in P_2O_5 content after treatment with acetic acid. This occurred despite the presence of significant amounts of rare earth elements. Moreover, studies have explored the potential interaction between acetic acid and rare earth elements, and relevant results are reported [11].

In the case of Guizhou (China), where calcium and magnesium carbonates comprise up to 50% of the ore, both single-factor and orthogonal experimental designs were applied using organic acids such as acetic, citric, and lactic acids [114]. Results from the single-factor tests suggested that only a solid-to-liquid ratio of 1:40 achieved complete dissolution of carbonate phases. The orthogonal experiment further revealed the following order of influence on enrichment effectiveness: acid type > reaction time > acid concentration > process temperature. These findings support the conclusion that acetic acid is most effective at lower concentrations and temperatures. Similar outcomes were observed in experiments on phosphorites from the Akash deposit in Iraq [115].

In Morocco's Khouribga phosphate deposit, single-factor studies demonstrated that when acetic acid concentration reached 7%, the P_2O_5 content increased from 18% to 30% [116].

In the Epirus phosphate region of Greece, researchers investigated a two-stage leaching process using acetic acid [117–119]. During the first stage, freely occurring calcite and dolomite compounds dissolved. In the second stage, carbonates bound within the phosphate matrix were successfully decomposed. These findings were validated through kinetic modeling. Furthermore, the acetic acid–carbonate reaction was found to produce unstable carbonic acid, which decomposes into water and carbon dioxide, a factor that provides control over the overall reaction rate.

In Uzbekistan, similar research has been conducted on the Jeroi-Syrdarya phosphate deposit [120]. When acetic acid was applied under a solid-to-liquid ratio of 1:8 at 25 °C, the initial P_2O_5 content increased from 15% to 22%. According to the authors, increasing the solid-to-liquid ratio proportionally increases the amount of acid solution required, highlighting the need for further research on safe disposal and neutralization of spent acids.

One of the main drawbacks of phosphate ore beneficiation using organic acids lies in the lack of established methods for the safe disposal of spent acid solutions. For this process to be economically viable, it is essential to develop strategies for recycling and reusing the organic acid used in the leaching process. In this context, researchers have proposed the regeneration of used organic acid solutions through the application of sulfuric acid [105, 106, 110, 111, 121].

Specifically, in the case of calcium carbonate interacting with acetic acid, the resulting calcium acetate solution can be treated with sulfuric acid to precipitate pure gypsum (calcium sulfate) free from phosphate contaminants. To achieve this, the sulfuric acid must be added in stoichiometric amounts sufficient to fully bind the calcium content. An excess of sulfuric acid, however, may introduce unwanted leaching of phosphate during reuse of the regenerated acid. Alternatively, some studies suggest regenerating the acid solution by introducing pressurized carbon dioxide gas to facilitate the separation process [111, 122].

The use of organic acids offers several advantages, including low environmental impact, minimal reactivity with phosphate components, the possibility of regenerating the spent acid, and the production of high-purity, clean end-products. Moreover, the water solubility of the resulting organic acid salts allows for easy removal from precipitates during post-processing [123–127]. Due to these advantages, organic acid-based beneficiation is increasingly viewed as one of the most promising and efficient methods for processing low-grade phosphate ores [128, 129].

Based on the conducted literature review, a comprehensive classification of global phosphate deposits and their reserves has been established, alongside a detailed overview of the principal phosphate raw material sites in Kazakhstan and their corresponding resource indicators. Furthermore, the characteristics of low-grade phosphorites and anthropogenic mineral formations have also been outlined.

A systematic analysis and generalization of the available data on beneficiation methods, approaches, and technologies for low-grade and waste-derived phosphate materials suggest that the application of organic acids represents a promising direction in this field. Among them, acetic acid has been shown to be particularly effective in the selective leaching of carbonate compounds from phosphate raw materials, with its efficacy in such processes well-supported by the literature.

Chapter 3
Experimental Section: Research Methods and Objects

3.1 Research Methods

The experimental part of the study was carried out in the laboratories of M. Auezov South Kazakhstan University, utilizing the available equipment base of two scientific centers: the "Structural and Biochemical Materials" Regional Engineering Testing Laboratory and the "Quality" Laboratory of Physico-Chemical Research Methods.

The experimental procedure for the beneficiation of low-grade phosphorites was divided into three major stages:

1. Preparation of raw material.
2. Acid leaching (beneficiation) process.
3. Compositional and structural analysis of the treated material.

3.1.1 Preparation of Raw Phosphate Samples

The preparation of raw phosphate rock involved mechanical grinding and particle size classification. A MSHL-1 laboratory ball mill was employed to crush the raw phosphate material. The steel drum of the mill was loaded with the sample and steel grinding balls, and the material was pulverized via the collision and abrasion mechanism during drum rotation. The resulting powder was then fractionated using a vibratory sieve of the "Analysette" brand, which ensured uniform particle size for further testing.

Y. Raiymbekov, *Enrichment of Low-Grade Phosphorites with Organic Acids*,
https://doi.org/10.1007/978-3-032-02910-2_3

3.1.2 Acid Beneficiation Procedure

For the acid-based beneficiation of low-grade phosphorites, an IKA C-MAG HS 7 magnetic stirrer was used. This equipment provided precise control over the heating temperature and stirring frequency of the reaction medium. A laboratory system was assembled wherein the IKA stirrer was coupled with an I-160MI laboratory ionometer for real-time monitoring of the pH level of the slurry during acid treatment. This allowed for optimal reaction conditions to be maintained throughout the leaching process (Fig. 3.1).

The objective of this stage was to investigate the selective removal of carbonate and dolomite components from phosphate ores using weak organic acids—specifically, acetic acid—as leaching agents.

For the experimental studies, finely powdered raw phosphate materials were used. Based on the thermodynamic data presented in Sect. 4.1, the temperature range for the acetic acid-based leaching process was selected as 293–313 K, with a leaching duration of 15–45 min. According to reactions (4.2)–(4.5), the complete decomposition of $CaCO_3$, $MgCO_3$, K_2O, and Al_2O_3 compounds present in the phosphate raw material was ensured by calculating the required amount of acetic acid and the necessary dilution volume based on its solubility in water. From these calculations, a liquid-to-solid ratio of 3:1 was established.

The diluted acetic acid solution used in the process had a density of $d = 1.030$, a pH value of 2.33, and a dissociation constant of $pK_a = 6.2 \times 10^{-2}$. The hydrogen ion concentration [H^+] was determined from the measured pH values using the formula:

$$\left[H^+\right] = 10^{-pH} \tag{3.1}$$

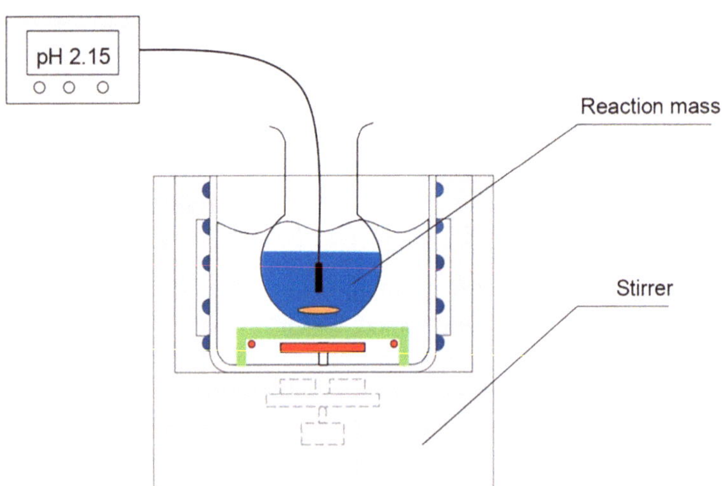

Fig. 3.1 Experimental setup

Based on the resulting [H^+] values, the concentration of the acid in the leaching solution (C, mol/L) was calculated. Subsequently, the consumption level of acetic acid in the leaching process (α) was determined using the following formula (3.2):

$$\alpha = \frac{(C_1 - C_2)}{C_1} \cdot 100, \tag{3.2}$$

where, C_1 is the concentration of acetic acid in the original solution (mol/L), and C_2 is the concentration of acetic acid in the post-leaching slurry (mol/L).

Using these values, the reaction rate was evaluated according to the following expression:

In the present study, the reaction rate (∂) for the acid leaching process was evaluated using the classical definition derived from stoichiometric and volumetric considerations (3.3):

$$\partial = \frac{\Delta n}{V \Delta \tau}, \tag{3.3}$$

where, Δn is the change in the number of moles of the reactant or product during the process, mol; V is the volume of the reaction mixture, L; $\Delta \tau$ is the duration of the process, min.

This expression provides a quantitative measure of the rate of the dissolution reaction under defined experimental conditions. It is particularly effective for systems where the amount of reactant consumed or product formed can be directly or indirectly measured through concentration, mass loss, or titrimetric analysis.

The efficiency of the leaching process was assessed based on the increase in P_2O_5 concentration (%). The degree of P_2O_5 enrichment (P_2O_5ED) was calculated using the following equation (3.4):

$$P_2O_{5\,ED} = \left[\frac{B \cdot b}{P \cdot p} \right] \cdot 100 \tag{3.4}$$

where, B is the mass of the enriched (leached) material (g), b is the P_2O_5 content in the enriched material (%), P is the mass of the initial raw material (g), p is the P_2O_5 content in the initial raw material (%).

To determine the overall enhancement of the valuable component in the concentrate, the "enrichment efficiency" (E) parameter was introduced, which serves as an index of process optimization. It was calculated using the following expression (3.5):

$$E = \frac{P_2O_{5\,ED} \cdot (100 - B)}{100 - A} \tag{3.5}$$

where, P_2O_5ED is the increase in P_2O_5 level (%), A is the P_2O_5 content in the raw material (%), B is the P_2O_5 content in the enriched material (%).

Additionally, the concentration ratio K, which indicates how many times the content of the valuable component increased in the concentrate compared to its initial concentration, was calculated as follows (3.6):

$$K = \frac{B}{A}, \tag{3.6}$$

where, A is the P_2O_5 content in the raw material (%), B is the P_2O_5 content in the enriched concentrate (%).

For leaching with citric and formic acids, the temperature range was extended to 25–70 °C with a process duration of 15–60 min. Based on the stoichiometric requirements for the complete dissolution of $CaCO_3$ and $MgCO_3$ compounds, the necessary quantities of citric and formic acids were calculated, along with the volume of water needed for dilution based on their aqueous solubility. These calculations resulted in a liquid-to-solid ratio of 3:1, and a more refined acid-to-solid ratio (A:S) of 2.5:1. The effectiveness of the leaching process was monitored using parameters such as CO_2 release, pH shift, and P_2O_5 content increase.

3.1.3 Analysis of Composition and Microstructure of Products

Comprehensive characterization of the solid products obtained after beneficiation was performed using classical chemical analysis methods, in accordance with national standards (GOST): GOST 21560.0-82, GOST 21560.1-82, GOST 4212-78, GOST 12083-78, GOST 25794-83, GOST 20851.2-75, GOST 20851.4-75, GOST 20851.3-93, and GOST 26318.3-84.

The quantitative analysis of carbonate content was performed through acid digestion followed by back titration. A representative powdered sample weighing approximately 0.2 g was transferred into a conical flask with a volume capacity of 100–150 mL. The sample was first moistened with distilled water to facilitate subsequent reactions. Then, using a burette or pipette, 25 mL of 0.2N hydrochloric acid (HCl) solution was added to the flask.

To minimize the loss of HCl and to ensure controlled reaction conditions, the flask was sealed with a stopper equipped with a vertical glass tube (3–5 mm in diameter and approximately 25 cm in length) functioning as a gas outlet. The flask was then heated gently on an electric hotplate until visible effervescence indicated the evolution of carbon dioxide gas (CO_2), confirming the decomposition of carbonate minerals. Care was taken to avoid boiling, as this would result in the volatilization of HCl, compromising the accuracy of the analysis.

Once decomposition was complete, the solution was brought to a brief boil, heating was discontinued, and all interior surfaces of the flask, including the stopper and glass tube, were rinsed with distilled water to recover any residual material. Upon cooling, three drops of 0.1% methyl red indicator solution were added to the solution. The excess acid was then titrated with 0.2N sodium hydroxide (NaOH) until the color changed from pink to yellow, indicating the endpoint.

Following the carbonate determination, the same solution was used for the analysis of calcium (Ca^{2+}) and magnesium (Mg^{2+}) ions. To eliminate any interfering substances such as iron oxides or organic matter, potassium permanganate ($KMnO_4$) solution was added dropwise until a persistent pink color was observed, indicating complete oxidation.

The entire contents of the flask were then transferred to a beaker, and the flask was rinsed with distilled water. Ammonia solution (NH_3) was added dropwise to the heated solution until a faint ammonia odor was detected, indicating slightly alkaline conditions. While maintaining continuous stirring, 25 mL of 0.2N ammonium oxalate ($(NH_4)_2C_2O_4$) solution was added from a burette to precipitate calcium oxalate.

The solution was subsequently heated to near boiling, then allowed to cool, transferred to a 200 mL volumetric flask, diluted to the mark with distilled water, and mixed thoroughly. The final solution was filtered through a dry filter paper into a clean, dry 100 mL volumetric flask. The first two fractions of the filtrate (10–15 mL each) were discarded to ensure analytical purity. Filtration was continued until exactly 100 mL of clear filtrate was collected, suitable for further analysis.

The phosphate content was determined via a wet digestion method using a mixed acid system. A 2.0 g portion of the sample was placed in a 250 mL beaker, followed by the addition of 15 mL of nitric acid (HNO_3) and 5 mL of hydrochloric acid (HCl). The mixture was gently heated to boiling and maintained under a watch glass cover until complete dissolution of the solid phase was achieved. After dissolution, the volume was adjusted to 50 mL by the addition of distilled water, and the mixture was boiled for an additional 5 min to ensure homogeneity.

The resulting solution was quantitatively transferred to a 250 mL volumetric flask, cooled to room temperature, and diluted to the mark with distilled water. After thorough mixing, the solution was filtered through dry filter paper into a clean, dry container. The initial portion of the filtrate was discarded to avoid contamination from adsorbed analytes or incomplete dissolution products.

In parallel, instrumental physico-chemical analysis was conducted using high-precision analytical instruments to determine the phase, elemental, and microstructural properties of the treated samples. The following equipment was employed:

1. JEOL JSM-6490 LV Scanning Electron Microscope (Japan)—Enabled morphological imaging of phosphate surfaces and mapping of elemental composition through Energy Dispersive Spectroscopy (EDS).
2. IR-Prestige 21 FTIR Spectrometer (Japan)—Covered the 7800–350 cm^{-1} range. Equipped with Michelson's interferometer and single-beam optics, this instrument allowed the identification of functional groups and bonding types. Spectra interpretation was supported by the IRsolution software.
3. NEOPHOT-21 Metallographic Microscope (Germany)—Used for visual identification and classification of mineral structures. STRUERS (Denmark) sample preparation equipment was employed to prepare thin polished sections.

4. Q-1500 D Derivatograph (Hungary)—Used to conduct Differential Thermal Analysis (DTA), revealing material dehydration, structural transitions, and thermal stability of phosphate components.
5. Bruker D8 X-ray Diffractometer (Germany)—Deployed at the Institute of Metallurgy and Ore Beneficiation (Satbayev University) for X-ray Diffraction (XRD) analysis to identify crystalline phases. Data interpretation was conducted using Diffrac Plus Search software, referencing the PDF2 diffraction pattern database (>450,000 compounds).

3.1.4 Thermodynamic Modeling

To evaluate the thermodynamic feasibility of the chemical reactions occurring during the beneficiation process, calculations were carried out using the HSC 9.3 software developed by the Finnish company Outokumpu. This program includes a dedicated Reaction Equation module, which enables the calculation of thermodynamic functions—such as enthalpy (ΔH), entropy (ΔS), Gibbs free energy (ΔG), and heat capacity (C_p)—for individual substances, as well as their changes during chemical reactions. These calculations are based on standard thermodynamic properties available in the software's integrated database.

For any given compound or reaction, the software uses the standard values at 298 K: standard enthalpy ($\Delta H_{(298)}$), standard entropy ($\Delta S_{(298)}$), molar heat capacity (C_p) to compute the values of thermodynamic functions at a user-specified temperature range.

The enthalpy at temperature T is calculated using the following expression (3.7):

$$\Delta H_T = H_{298} + \int_{298}^{T} C_p \cdot \Delta T + \sum H_\Phi, \tag{3.7}$$

where, H_{298} is the standard enthalpy of the selected compound at 298 K; C_p is the molar heat capacity; H_Φ *represents the enthalpy associated with phase transitions.*

The entropy at temperature T is determined from (3.8):

$$\Delta S_T = S_{298} + \int_{298}^{T} \frac{C_p}{T} \cdot \Delta T + \frac{\sum H_\Phi}{T}, \tag{3.8}$$

where, S_{298} is the standard entropy at 298 K; C_p is the molar heat capacity; $\frac{\sum H_\Phi}{T}$ accounts for the entropy of phase transitions.

Once the enthalpy and entropy at a given temperature are determined, the Gibbs free energy is calculated via (3.9):

$$\Delta G_T = \Delta H_T - T \cdot \Delta S_T. \tag{3.9}$$

Calculating the changes in thermodynamic functions during chemical reactions provides critical insights into the reaction's direction and feasibility. The sign of ΔG indicates whether the process can proceed spontaneously in the given direction. The value of ΔH determines the heat effect—whether the process is exothermic or endothermic—and thus whether it releases or absorbs energy from the surroundings. The equilibrium constant (K) is related to ΔG and defines the ratio between the concentrations of products and reactants at equilibrium.

For heterogeneous reactions, the HSC Reaction Equation module computes the changes in ΔH, ΔS, and ΔG based on the following formulas (3.10)–(3.12):

$$\Delta H_T = \sum s_i H_{i(\text{product})} - \sum s_i H_{i(\text{reactant})}, \tag{3.10}$$

$$\Delta S_T = \sum s_i S_{i(\text{product})} - \sum s_i S_{i(\text{reactant})}, \tag{3.11}$$

$$\Delta G_T = \sum s_i G_{i(\text{product})} - \sum s_i G_{i(\text{reactant})}, \tag{3.12}$$

where, s_i is the stoichiometric coefficient of the respective species.

The equilibrium constant (K) of a heterogeneous reaction is directly related to ΔG and is computed using (3.13):

$$\ln K = -\frac{\Delta G_T}{RT}, \tag{3.13}$$

where, R is the universal gas constant; T is the absolute temperature (in Kelvin).

The HSC 9.3 database contains standard thermodynamic values—enthalpy, entropy, and Gibbs free energy—for over 17,000 inorganic and organic compounds. These values can be employed to conduct detailed thermodynamic analyses over a user-defined temperature range for a specified reaction system.

3.1.5 Kinetic Modeling and Rate Determination of the Acidic Leaching of Low-Grade Phosphorites

The rate of chemical reaction is a fundamental parameter for understanding and modeling heterogeneous solid–liquid processes, such as the leaching of low-grade phosphorites with organic acids. The rate of reaction characterizes the transformation of reactants into products per unit time and is crucial for optimizing process efficiency and designing scaled-up industrial reactors.

To analyze the kinetic behavior of the acetic acid leaching process, the shrinking core model (SCM) was adopted, which is widely applied in heterogeneous solid–liquid systems where reaction proceeds through the shrinking of an unreacted solid core. Under the assumption that the reaction is controlled by product layer diffusion or surface chemical reaction, the data were linearized using the following empirical expression (3.14):

$$1-(1-\alpha)^{1/3}=k\cdot\sqrt{\tau} \tag{3.14}$$

where, α is the fractional conversion of phosphate; k is the apparent rate constant ($min^{-1/2}$); τ is the leaching time, min.

This formulation corresponds to a mixed diffusion-reaction control mechanism, which is often observed in porous solids such as sedimentary phosphorites where both mass transfer and surface kinetics are rate-limiting at different stages of the process.

To estimate the apparent activation energy (E_a) of the leaching reaction, the Arrhenius equation was used in its logarithmic linear form (3.15):

$$\ln k = f\left(\frac{1}{T}\right)\rightarrow \ln k=\ln A-\frac{E_a}{RT}. \tag{3.15}$$

A linear plot of ln k versus $1/T$ was constructed based on kinetic data obtained at different temperatures. The slope of the regression line tan φ (i.e., the tangent of the angle formed with the abscissa) was used to compute the activation energy according to the following expression (3.16):

$$E_0=8.314\cdot\mathrm{tg}_{\Phi}\left(\mathrm{kJ/mol}\right) \tag{3.16}$$

where, R = 8.314 J/mol K is the universal gas constant; tan φ is the slope of the linear dependence ln k vs. $1/T$.

The activation energy is a crucial thermokinetic parameter that reflects the sensitivity of the leaching rate to temperature variations and provides insights into the underlying reaction mechanism. Lower activation energies typically suggest diffusion-controlled processes, while higher values are indicative of chemically controlled reactions.

Statistical Analysis

All experimental data, including kinetic constants and activation energies, were subjected to comprehensive statistical processing using the Online Statistics Application [130]. The software enabled the construction of regression curves, evaluation of correlation coefficients, and error propagation analysis to ensure the reliability and reproducibility of the obtained kinetic parameters.

This methodological approach ensured a rigorous kinetic interpretation of the acid leaching process and allowed for the mechanistic validation of the acetic acid treatment applied to carbonate-rich, low-grade phosphorite ores.

Table 3.1 Chemical composition of phosphate-siliceous layered rocks from the Zhanatas deposit

Component	Content, %
Total P_2O_5	12.07
CaO	22.73
MgO	2.17
K_2O	1.15
Al_2O_3	5.83
Fe_2O_3	2.52
SiO_2	37.78

3.2 Research Objects

3.2.1 *Phosphate-Silica Layer Rocks of the Zhanatas Deposit*

For the purpose of this research, representative samples of phosphate-siliceous layered rocks of carbonate-type origin were collected from the Zhanatas phosphate deposit, located in the Karatau phosphorite basin. These samples represent technogenic sedimentary formations that have accumulated as a result of long-term open-pit mining activities. The chemical composition of the samples was determined through standard analytical techniques and is summarized in Table 3.1.

As shown in Table 3.1, the primary valuable component in the studied samples is total phosphorus pentoxide (P_2O_5), which accounts for 12.07% by mass. This value reflects the material's low-grade phosphorite classification. In addition to P_2O_5, the samples exhibit a high content of calcium oxide (CaO)—22.73%—indicative of the presence of carbonate minerals such as calcite and dolomite. Magnesium oxide (MgO) is present at 2.17%, which also supports the carbonate classification.

The samples also show elevated concentrations of aluminum oxide (Al_2O_3, 5.83%) and iron oxide (Fe_2O_3, 2.52%), suggesting the existence of clay minerals, ferruginous components, and possible associations with aluminosilicate phases. The significant presence of silicon dioxide (SiO_2, 37.78%) reflects the predominance of quartz and other siliceous phases within the mineral matrix.

To gain a deeper understanding of the microstructure and elemental distribution within these rocks, scanning electron microscopy (SEM) imaging and energy-dispersive spectroscopy (EDS) analysis were conducted. SEM micrographs at 40× and 800× magnification revealed the morphological characteristics of the mineral phases, while EDS provided semi-quantitative data on the elemental-weight composition of the examined areas. The corresponding results are presented in Fig. 3.2 and Table 3.2, respectively.

These analytical findings form the basis for evaluating the beneficiation potential of the Zhanatas phosphate-siliceous rocks using selective acid leaching techniques and support further thermodynamic and kinetic investigations.

To investigate the microstructural and elemental characteristics of phosphate-siliceous layered rocks obtained from the Zhanatas deposit, a scanning electron microscope (SEM) combined with energy-dispersive spectroscopy (EDS) analysis

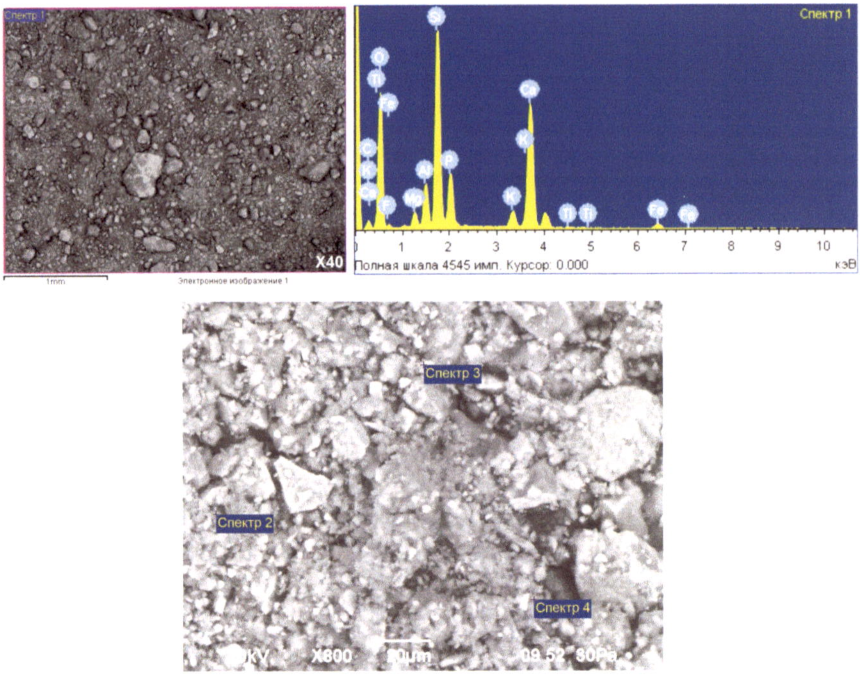

Fig. 3.2 SEM micrographs and EDS spectral analysis results of phosphate-siliceous layered rocks from the Zhanatas Deposit

was employed. Figure 3.2 presents representative SEM micrographs of the studied sample at magnifications of 40× and 800×, revealing the heterogeneity of the rock matrix and the presence of well-defined grain boundaries, typical of sedimentary phosphorites.

The EDS analysis was performed on four discrete spectra (Spectra 1–4) across different micro-regions of the sample. The elemental-weight composition (wt%) and its corresponding oxide recalculation are summarized in Table 2.3.

The spectral data confirm the dominance of calcium and silicon-containing compounds, with calcium oxide (CaO) accounting for an average of 21.18%, and silicon dioxide (SiO_2) reaching 32.51%. Notably, phosphorus pentoxide (P_2O_5) was determined to be 12.11%, supporting the classification of these materials as low-grade phosphorites with a significant siliceous and carbonate component.

The elevated concentrations of alumina (Al_2O_3) and ferric oxide (Fe_2O_3) further indicate the presence of aluminosilicate and ferruginous phases, which are typically resistant to conventional beneficiation techniques. The presence of fluorine (F) in trace amounts suggests the occurrence of fluorapatite phases.

In order to complement the morphological and elemental analyses, Fourier-transform infrared spectroscopy (FTIR) was employed to characterize the structural-functional groups and mineral associations present in the sample. The IR spectroscopic data are provided in Fig. 3.3 and detailed in Table 3.3.

Table 3.2 Elemental and oxide-recalculated composition of phosphate-siliceous layered rocks (Zhanatas deposit)

	Spectrum 1		Spectrum 2		Spectrum 3		Spectrum 4		Average	
Element	wt% (element)	wt% (oxide)	wt% (element)	wt% (oxide)	wt% (element)		wt% (element)	wt% (oxide)	wt% (element)	wt% (oxide)
C	7.94	–	5.18	–	2.03	–	5.71	–	5.21	–
O	46.20	–	49.71	–	55.29	–	52.39	–	50.89	–
F	1.06	–	0.74	–	–	–	–	–	0.9	–
Mg	1.28	2.12	0.91	1.50	1.46	2.42	1.15	1.90	1.20	1.98
Al	3.18	6.01	3.10	5.85	2.74	5.17	3.42	6.46	3.11	5.87
Si	14.40	30.80	15.57	33.30	16.83	35.99	14.01	29.96	15.20	32.51
P	5.21	11.93	4.98	11.40	5.07	11.61	5.91	13.53	5.29	12.11
K	1.82	2.19	1.19	1.43	1.60	1.92	0.49	0.59	1.27	1.53
Ca	15.77	22.06	16.28	22.77	13.09	18.31	15.42	21.57	15.14	21.18
Fe	1.80	2.57	2.33	3.33	1.77	2.53	1.42	2.03	1.83	2.61
Ti	0.22	0.36	–	–	0.12	0.20	0.07	0.11	0.13	0.21

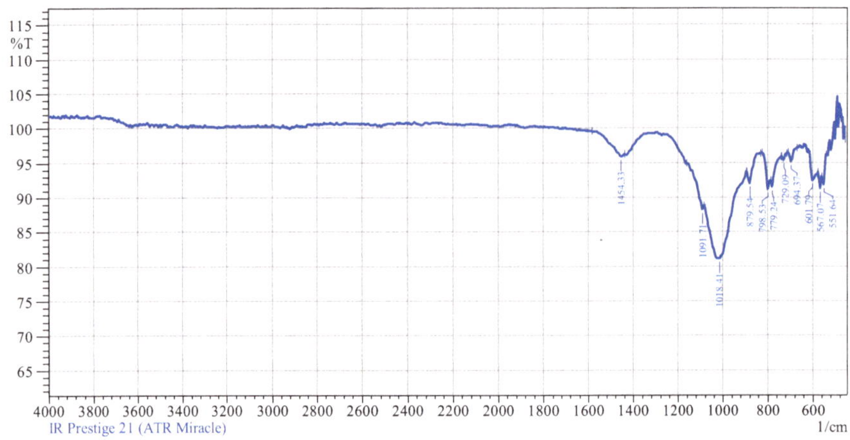

Fig. 3.3 FTIR spectra of phosphate-siliceous layered rocks

Table 3.3 Peak positions of FTIR spectra

Peak	Intensity	Corr. intensity	Base (*H*)	Base (*L*)	Area	Corr. area
551.64	91.789	2.438	559.36	540.07	0.577	0.105
567.07	91.309	2.019	574.79	559.36	0.534	0.071
601.79	92.428	2.794	613.36	578.64	1.079	0.239
694.37	95.213	1.524	709.80	671.23	0.641	0.100
729.09	95.427	0.651	736.81	713.66	0.417	0.027
779.24	91.553	1.517	786.96	740.67	1.217	0.077
798.53	91.152	2.562	825.53	786.96	1.057	0.108
879.54	92.036	2.145	891.11	829.39	1.450	0.158
1018.41	81.130	9.065	1080.14	894.97	11.705	4.203
1091.71	88.277	1.094	1161.15	1083.99	2.793	−0.077
1454.33	95.915	0.654	1585.49	1438.90	1.361	0.080

The infrared spectral analysis of the phosphate-siliceous layered rock samples from the Zhanatas deposit revealed the presence of multiple characteristic absorption bands, which were used to identify the functional groups and mineral phases within the material:

- 550–570 cm^{-1}: This range is attributed to P–F (phosphorus-fluorine) vibrational modes, indicating the presence of fluorapatite-like compounds commonly observed in sedimentary phosphorites [131].
- 600–780 cm^{-1}: The absorption in this region corresponds to aluminate-type compounds, typically associated with aluminosilicate minerals, such as kaolinite or gibbsite components in the phosphate matrix [132].
- 800–960 cm^{-1}: These wave numbers are indicative of the Si–O–Si asymmetric stretching vibrations, confirming the presence of a silicate framework within the

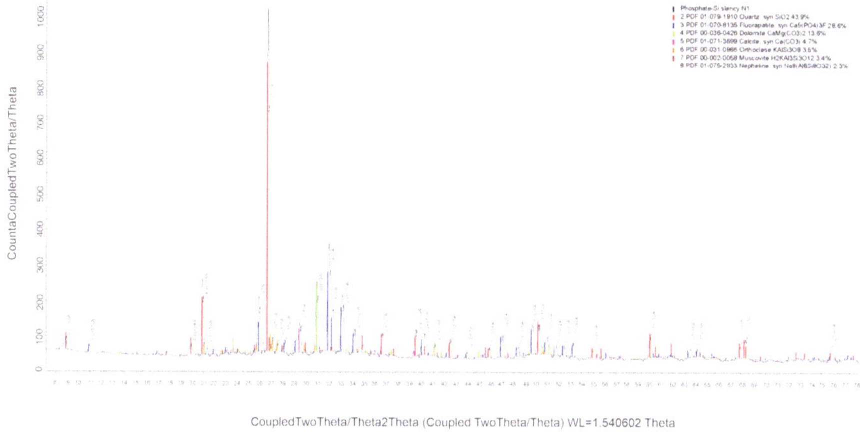

Fig. 3.4 X-ray diffraction (XRD) pattern of phosphate-siliceous layered rocks

phosphate matrix. This reflects the structural integrity of siliceous phases embedded in the phosphorite body [133].

- Additionally, the absorption band observed between 1000 and 1060 cm^{-1}, which is also characteristic of Si–O–Si bonds, further supports the existence of both layered and chain silicate structures within the mineral framework of the raw material. This dual occurrence signifies a heterogeneous silicate composition that could influence the reactivity and beneficiation behavior of the ore.
- A distinct and sharp band observed in the 1435–1450 cm^{-1} range is associated with carbonate (CO_3^{2-}) vibrational modes, confirming the presence of carbonate phases, such as calcite or dolomite, within the phosphate-siliceous rocks. These carbonates represent the principal impurities in low-grade phosphorites and play a key role in determining the choice of beneficiation method [134].

The X-ray diffraction (XRD) analysis of the phosphate-siliceous layered rocks was performed using a Bruker D8 Advance diffractometer, a high-precision instrument that allows for detailed phase identification of complex geological samples. The results of the XRD study are presented in Fig. 3.4 and summarized in Table 3.4, highlighting the crystalline phases identified in the Zhanatas phosphate ores. These findings corroborate the FTIR data and provide structural insight into the mineralogical composition of the studied materials.

The XRD analysis of the phosphate-siliceous layered rock samples revealed a dominant presence of quartz (SiO_2), accounting for 43.9% of the total crystalline content. Fluorapatite ($Ca_5(PO_4)_3F$), the primary phosphate-bearing phase, was found in significant quantity (28.6%), confirming the phosphatic nature of the deposit. Additionally, dolomite ($CaMg(CO_3)_2$) and calcite ($CaCO_3$) were identified with proportions of 13.6% and 4.7%, respectively, indicating the presence of substantial carbonate content.

Table 3.4 Mineral phases identified in phosphate-siliceous layered rocks based on XRD analysis

PDF reference	Identified phase	Content, %
PDF 01-079-1910	Quartz, syn SiO_2	43.9
PDF 01-070-8135	Fluorapatite, syn $Ca_5(PO_4)_3F$	28.6
PDF 00-036-0426	Dolomite, $CaMg(CO_3)_2$	13.6
PDF 01-071-3699	Calcite, syn $CaCO_3$	4.7
PDF 00-031-0966	Orthoclase, $KAlSi_3O_8$	3.5
PDF 00-002-0058	Muscovite, $H_2KAl_3Si_3O_{12}$	3.4
PDF 01-075-2933	Nepheline, syn $Na_8(Al_8Si_8O_{32})$	2.3

Minor mineral phases include orthoclase ($KAlSi_3O_8$, 3.5%) and muscovite ($H_2KAl_3Si_3O_{12}$, 3.4%), both of which are aluminosilicates commonly encountered in sedimentary and metamorphic environments. A small fraction of nepheline ($Na_8(Al_8Si_8O_{32})$, 2.3%) was also detected, suggesting the occurrence of sodium-rich feldspathoid components.

The differential thermal analysis (DTA) of phosphate-siliceous layered rocks was carried out using the Q-1500 D derivatograph, with the heating process conducted up to a maximum temperature of 1000 °C and a baseline leveling degree of 5 (Fig. 3.5). Throughout the thermal experiment, several key transitions were observed that reflect the structural and phase transformations occurring in the material.

At lower temperatures, specifically in the range of 60–150 °C, the thermal curve exhibited a mild endothermic signal, which corresponds to the release of physically adsorbed and structural water from the sample. As the temperature increased to approximately 200–270 °C, the transformation of quartz from its low-temperature to high-temperature cubic modification was initiated, indicating a phase rearrangement within the siliceous framework of the rock.

Further heating to the range of 350–585 °C resulted in the thermal decomposition of siderite ($FeCO_3$), followed by the oxidation of divalent iron to trivalent iron (Fe^{2+} to Fe^{3+}) in the form of Fe_2O_3 between 600 and 830 °C. This transformation reflects the progressive destabilization of iron-bearing phases under oxidizing thermal conditions.

Between 730 and 790 °C, a pronounced endothermic response was recorded, corresponding to the breakdown of dolomite into its constituent carbonates—calcium carbonate ($CaCO_3$) and magnesium carbonate ($MgCO_3$). As heating continued to the range of 830–940 °C, these carbonate phases further decomposed, releasing carbon dioxide and contributing to mass loss in the sample.

Structural disintegration of muscovite was observed between 850 and 900 °C, characterized by the collapse of its layered silicate framework. Simultaneously, in the range of 860–920 °C, calcite underwent complete dissociation. These overlapping events suggest complex interactions between aluminosilicate and carbonate minerals under thermal stress.

Overall, the duration of the thermal experiment was 88 min, during which the total mass loss of the sample amounted to 9.6%. This loss reflects a combination of

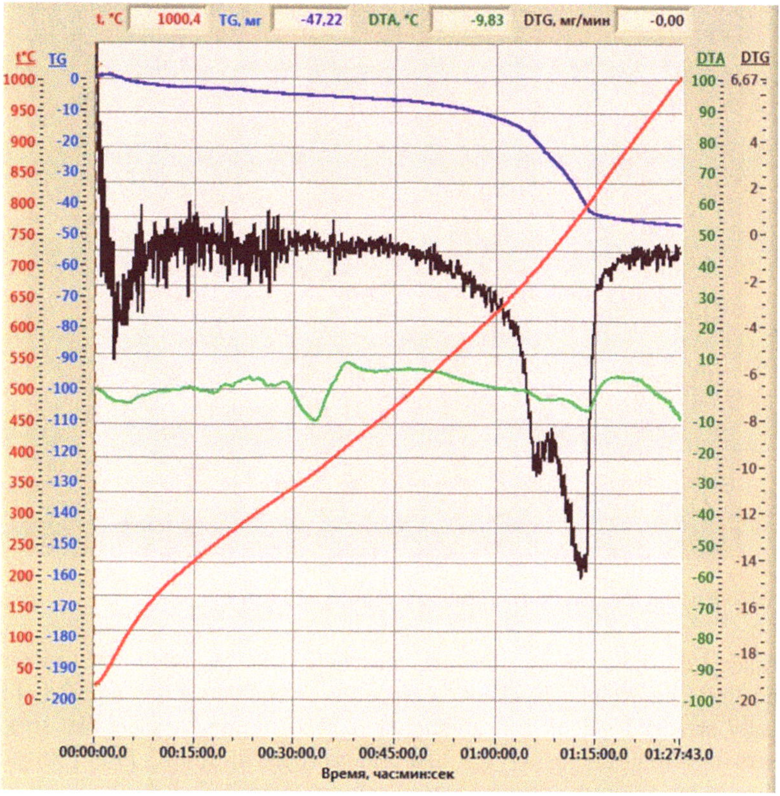

Fig. 3.5 Derivatogram of phosphate-siliceous layered rocks

dehydration, decarbonation, and dehydroxylation processes, confirming the mineralogical complexity and multi-phase composition of the studied phosphate-siliceous rocks.

The mineralogical composition of the studied sample was further investigated using a NEOPHOT-21 metallographic microscope. The optical micrograph of the phosphate-siliceous layered rocks (Fig. 3.6) revealed a poorly consolidated aggregate structure, where the dominant phase is a dark-brown hexagonal syngony phosphorite. According to the Mohs hardness scale, this mineral has a hardness of approximately 5 and a density of 3.1–3.2 g/cm^3.

The presence of siderite ($FeCO_3$), a mineral with a trigonal syngony and characteristic stepped cleavage, was also confirmed. Siderite is commonly associated with carbonate-bearing formations and contributes to the total iron content. Additionally, the carbonate group minerals include calcite and dolomite. Calcite is present as grainy aggregates with elongated rhombohedral crystal forms, having a Mohs hardness of 3 and a density ranging from 2.6 to 2.8 g/cm^3. Dolomite, on the other hand, occurs in a more porous, crystalline-granular morphology, exhibiting good cleavage, a Mohs hardness of 3.5–4, and a density of 2.86 g/cm^3.

Fig. 3.6 Optical micrograph of phosphate-siliceous layered rocks

Among the mica group minerals, muscovite was identified. This mineral is monoclinic and exhibits excellent cleavage. Its crystal lattice is composed of tetrahedrally arranged silicon and aluminum oxides, forming a sheet-like structure. Muscovite's Mohs hardness is 2.5, and its density ranges from 2.77 to 2.88 g/cm^3.

Finally, a significant portion of the quartz group mineral chalcedony was identified. Chalcedony is a cryptocrystalline variety of quartz with a fibrous microstructure. It typically fills cavities within magmatic rocks and has a compact and dense appearance. Chalcedony exhibits a Mohs hardness of 6.5–7 and a density of 2.58–2.64 g/cm^3.

3.2.2 Low-Grade Phosphorites of the Akzhar Deposit

The chemical composition of low-grade phosphorite samples obtained from the Akzhar deposit was determined through comprehensive chemical analysis. The results are presented in Table 3.5.

According to the results of the chemical analysis, the phosphate raw material contains 14.51% of phosphorus pentoxide (P_2O_5), indicating a moderate phosphate content typical of low-grade phosphorites. The dominant major oxide is calcium oxide (CaO), with a concentration of 31.43%, while magnesium oxide (MgO) is present at 2.08%. In addition to these major components, the presence of minor oxides such as aluminum oxide (Al_2O_3) and iron oxide (Fe_2O_3) was confirmed in

Table 3.5 Chemical composition of low-grade phosphorites from the Akzhar deposit

Component	Content, %
Total P_2O_5	14.51
CaO	31.43
MgO	2.08
K_2O	0.39
Al_2O_3	1.16
Fe_2O_3	0.51
SiO_2	20.08

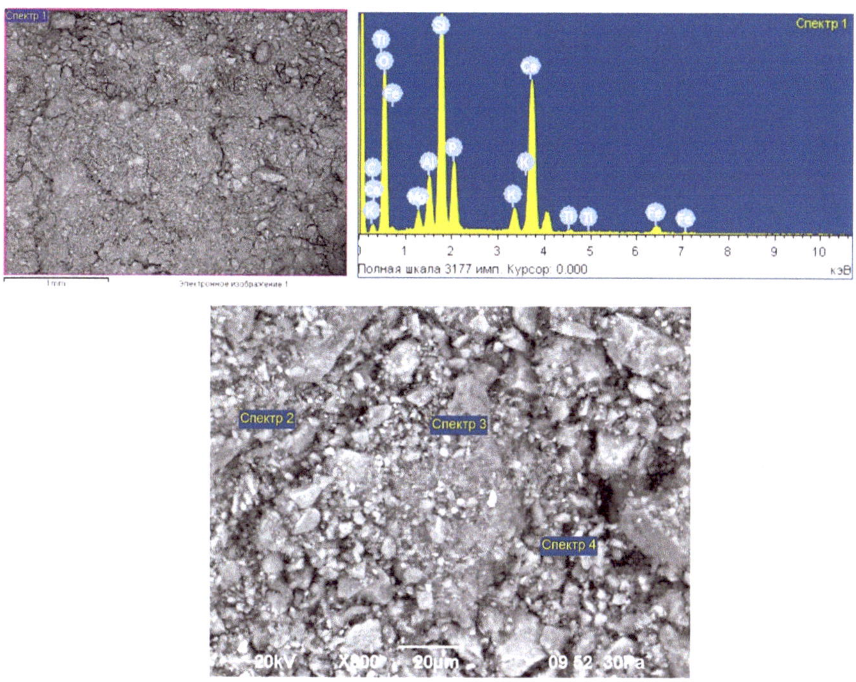

Fig. 3.7 SEM micrographs and EDS spectral analysis results of low-grade phosphorites from the Akzhar deposit

concentrations of 1.16% and 0.51%, respectively. These oxides may influence the reactivity and beneficiation behavior of the raw material.

The mineralogical and elemental microstructure of the Akzhar phosphorites was further investigated using scanning electron microscopy. Micrographs taken at 40× and 800× magnifications are provided in Fig. 3.7. Furthermore, the elemental-weight composition was assessed using energy-dispersive X-ray spectroscopy (EDS) for four different spectra. The results of this analysis are compiled in Table 3.6.

The results of energy-dispersive spectroscopy (EDS) conducted on four spectra from the low-grade phosphorite samples collected at the Akzhar deposit are summarized in Table 3.6. The micrographs were captured using a scanning electron microscope at magnifications of 40× and 800×.

Table 3.6 Elemental and oxide-recalculated composition of low-grade phosphorites from the Akzhar deposit

	Spectrum 1		Spectrum 2		Spectrum 3		Spectrum 4		Average	
Element	wt% (element)	wt% (oxide)	wt% (element)	wt% (oxide)	wt% (element)	wt% (oxide)	wt% (element)	wt% (oxide)	wt% (element)	wt% (oxide)
C	8.91	–	5.76	–	7.13	–	6.75	–	7.13	–
O	48.66	–	54.17	–	54.40	–	49.61	–	51.71	–
F	–	–	–	–	1.62	–	0.54	–	1.08	–
Mg	2.55	4.22	1.43	2.37	1.89	3.13	2.01	3.33	1.97	3.26
Al	0.46	0.86	0.71	1.34	1.07	2.02	0.12	0.22	0.59	1.11
Si	8.65	18.50	9.28	19.84	8.91	19.05	8.73	18.67	8.89	19.01
P	6.47	14.82	7.13	16.33	5.61	12.85	4.78	10.95	5.99	13.72
K	0.23	0.27	0.17	0.20	0.74	0.89	0.33	0.39	0.36	0.43
Ca	23.66	33.10	20.76	29.04	18.17	25.41	26.40	36.93	22.24	31.11
Fe	0.38	0.54	0.59	0.84	0.46	0.65	0.73	1.04	0.54	0.77
Ti	0.03	0.05	–	–	–	–	–	–	0.03	0.05

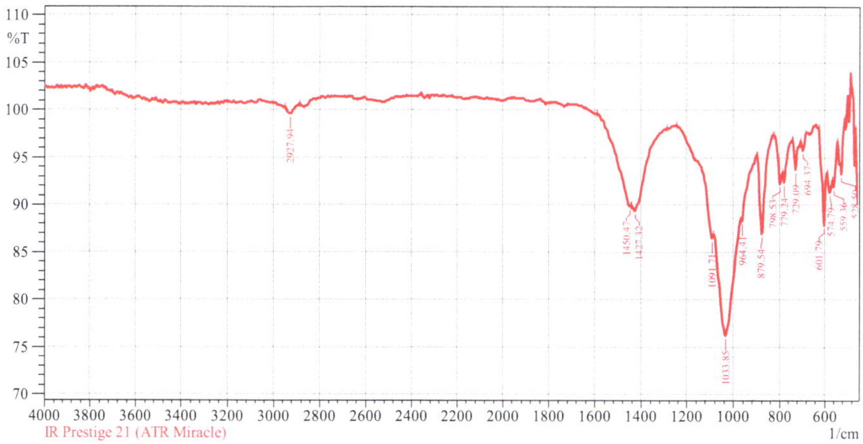

Fig. 3.8 FTIR spectra of low-grade phosphorites from the Akzhar deposit

Analysis of the EDS data indicates that the primary constituents of the phosphate raw material are calcium, phosphorus, and silicon, predominantly in the form of their respective oxides. The average oxide content across all four spectra reveals 31.11% CaO, 13.72% P_2O_5, and 19.01% SiO_2, confirming that the phosphate mineral phase is accompanied by a significant amount of silicate material. These values clearly classify the Akzhar samples as low-grade phosphorites based on the moderate concentration of phosphorus pentoxide.

In addition, magnesium oxide (MgO) is present at an average of 3.26%, while potassium oxide (K_2O) contributes 0.43%. The alumina (Al_2O_3) and ferric oxide (Fe_2O_3) contents are relatively minor, with average concentrations of 1.11% and 0.77%, respectively. Fluorine (F), although not detected in all spectra, was observed in spectra 3 and 4 with an average of 1.08%, indicating the probable presence of fluorapatite in the mineral assemblage. Titanium (Ti) was detected only in spectrum 1, with a concentration of 0.03%, suggesting its occurrence is minimal and sporadic in the studied samples.

These compositional insights provide crucial information for evaluating the beneficiation potential of the Akzhar phosphate raw material. The relatively low P_2O_5 content and significant silicate association imply a need for selective leaching or flotation techniques to achieve enrichment.

The sample was further characterized by Fourier-transform infrared spectroscopy (FTIR) to investigate the structural and bonding environment of mineral phases. The results of this analysis are presented in Fig. 3.8 and Table 3.7.

The FTIR spectral analysis of the Akzhar deposit's low-grade phosphorite samples revealed several characteristic absorption bands indicative of the mineralogical and chemical structure of the material. The absorption bands observed in the 400–500 and 800–960 cm^{-1} ranges correspond to the vibrational modes of Si–O–Si bonds, confirming the presence of silicate frameworks within the phosphate matrix.

Table 3.7 Peak positions of FTIR spectra

Peak	Intensity	Corr. intensity	Base (*H*)	Base (*L*)	Area	Corr. area
455.20	90.145	4.472	462.92	447.49	0.595	0.159
528.50	93.260	4.651	543.93	513.07	0.601	0.338
559.36	91.907	2.136	567.07	547.78	0.559	0.106
574.79	91.353	1.345	586.36	567.07	0.711	0.076
601.79	87.874	7.127	624.94	590.22	1.235	0.540
694.37	95.813	1.320	705.95	682.80	0.359	0.073
729.09	93.787	2.912	740.67	709.80	0.616	0.158
779.21	92.390	1.672	786.96	744.52	0.949	0.087
798.53	92.226	2.389	825.53	786.96	0.905	0.112
879.54	86.966	8.850	898.83	829.39	2.152	1.029
964.41	88.310	0.410	968.27	902.69	2.205	0.007
1033.85	76.294	11.224	1083.99	972.12	9.942	3.517
1091.71	86.493	0.941	1246.02	1083.99	4.237	−1.281
1427.32	89.387	1.358	1442.75	1288.45	3.823	0.210
1450.47	89.854	0.359	1589.34	1446.61	2.951	−0.456
2927.94	99.586	0.840	2951.09	2885.51	−0.015	0.121

The peaks in the 550–570 cm^{-1} region are attributed to P–F stretching vibrations, which are typically associated with fluorapatite minerals.

Furthermore, absorption bands in the 600–780 cm^{-1} region suggest the presence of aluminosilicate phases, while those between 1000 and 1060 cm^{-1} correspond to layered and chain-like silicate structures, also linked to Si–O–Si vibrations. The bands in the 1435–1450 cm^{-1} range are characteristic of CO_3^{2-} groups and indicate the presence of carbonate phases within the phosphate samples. Additionally, absorption peaks observed in the 2750–3000 cm^{-1} region are associated with organic compounds, possibly residues from biological matter or naturally occurring humic substances incorporated into the rock matrix during diagenesis [131–134].

The spectral data collectively suggest that the phosphate matrix of the Akzhar samples contains a complex mixture of silicate, carbonate, fluorinated, and minor organic components, each contributing to the infrared fingerprint of the material. These findings are crucial for understanding the physicochemical behavior of the phosphate during beneficiation processes such as acid leaching, thermal decomposition, or flotation.

To complement the FTIR findings and gain insight into the phase assemblage of the raw phosphate material, X-ray diffraction (XRD) analysis was conducted. The phase composition results are presented in Fig. 3.9 and detailed in Table 3.8.

According to the results of the X-ray diffraction (XRD) analysis, the mineralogical composition of low-grade phosphorite samples from the Akzhar deposit is predominantly represented by fluorapatite ($Ca_5(PO_4)_3F$), accounting for 48.76% of the total phase content. This phase is the principal phosphate-bearing mineral in the studied sample and is responsible for the phosphorus enrichment potential of the ore. Quartz (SiO_2) was the second most abundant component, with a content of

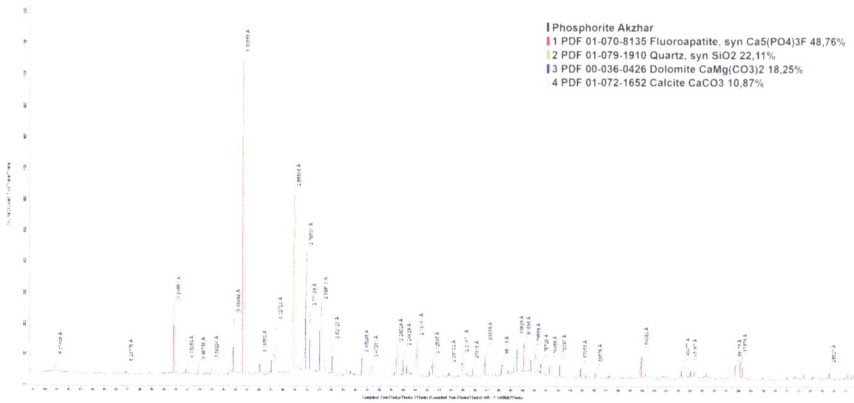

Fig. 3.9 X-ray diffraction (XRD) pattern of Akzhar deposit's low-grade phosphorite

Table 3.8 Mineral phases identified in low-grade phosphorites from the Akzhar deposit based on XRD analysis

PDF reference	Identified phase	Content, %
PDF 01-070-8135	Fluorapatite, syn $Ca_5(PO_4)_3F$	48.76
PDF 01-079-1910	Quartz, syn SiO_2	22.11
PDF 00-036-0426	Dolomite, $CaMg(CO_3)_2$	18.25
PDF 01-071-3699	Calcite, $CaCO_3$	10.87

22.11%, indicating a substantial presence of siliceous material in the phosphorite matrix.

Carbonate minerals were also prominent in the sample, with dolomite ($CaMg(CO_3)_2$) contributing 18.25% and calcite ($CaCO_3$) representing 10.87%. The presence of these carbonates is critical, as they influence the chemical behavior of the phosphorites during beneficiation, particularly in acid leaching processes. These findings confirm the complex multiphase nature of the Akzhar ore and underscore the need for selective chemical treatment strategies during processing.

To further investigate the thermal behavior of the sample, a differential thermal analysis (DTA) was conducted using a Q-1500 D derivatograph. The maximum programmed heating temperature reached 1000 °C. The results of this analysis, including thermal events associated with structural transformations, decomposition of carbonate phases, and loss of structural water, are presented in Fig. 3.10.

The differential thermal analysis (DTA) of low-grade phosphorite samples from the Akzhar deposit was performed using a Q-1500 D derivatograph under a controlled heating regime up to 1000 °C. The analysis provided valuable insights into the thermal behavior and compositional evolution of the sample during progressive heating.

The first thermal event, observed in the range of 90–170 °C, corresponds to endothermic effects associated with the release of adsorbed (physically bound)

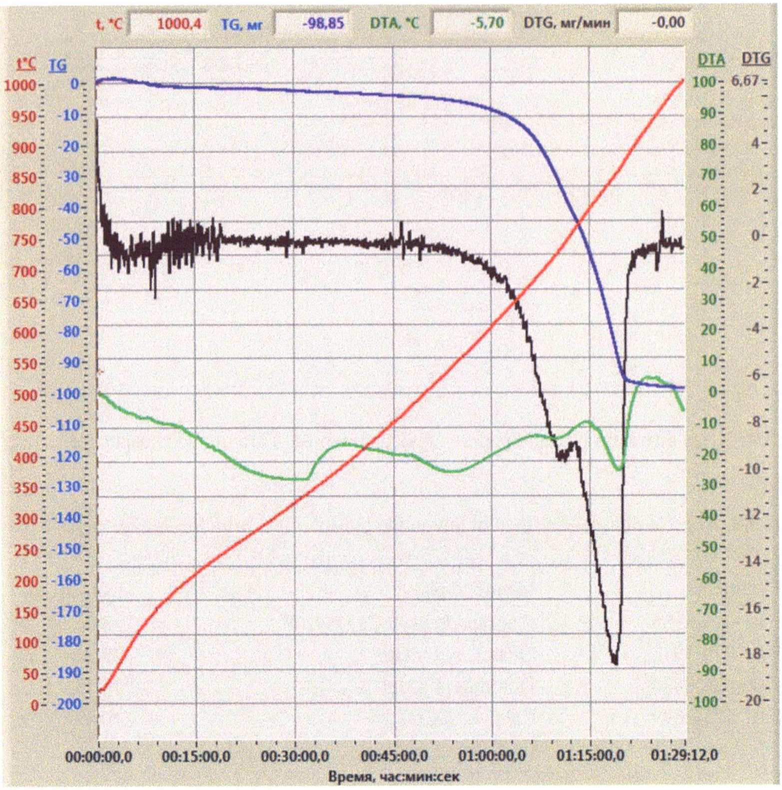

Fig. 3.10 Derivatogram of low-grade phosphorites from the Akzhar deposit

water from the sample surface and pores. As the temperature increases beyond 200 °C, a polymorphic transformation of quartz to its cristobalite form takes place, indicating structural rearrangements in the siliceous components.

The interval between 520 and 570 °C reveals further endothermic transitions, which are attributed to the dehydration of clay minerals within the sample matrix. This suggests the presence of hydrous aluminosilicates such as kaolinite or smectite. At temperatures above 700 °C, prominent decomposition of carbonate phases occurs, including dolomite ($CaMg(CO_3)_2$) and calcite ($CaCO_3$), releasing carbon dioxide and altering the mineral structure.

In the range of 940–980 °C, distinct exothermic peaks were recorded, indicating crystallization processes involving aluminous phases. These thermal effects imply structural rearrangement and possible formation of stable alumina-containing compounds during high-temperature treatment [135].

Overall, the duration of the thermal analysis was 90 min, during which the total mass loss of the sample reached 8.44%, confirming the elimination of volatile components and thermally unstable mineral phases.

To complement the thermal analysis, the mineralogical microstructure of the Akzhar phosphorite was further investigated. The raw sample was first subjected to mechanical preparation using a Struers sample preparation system to ensure high-quality polished cross-sections. Microscopic imaging was then carried out using a metallographic microscope, and the results are presented in Fig. 3.11. These images provide additional morphological evidence of the mineralogical phases observed during the thermal decomposition analysis.

The mineralogical micrographs of the studied phosphorite sample reveal that the dominant mineral phases are white-lustrous calcite and yellowish dolomite. These minerals typically occur in fine-grained aggregates characteristic of sedimentary and weakly metamorphosed lithologies. Both calcite and dolomite exhibit rhombohedral cleavage along three crystallographic directions, resulting in a well-developed mineralogical intergrowth pattern. Due to their similar optical and structural properties, distinguishing between the two minerals is often challenging. However, dolomite is known to exhibit higher Mohs hardness and greater specific gravity compared to calcite. According to the literature, dolomite's hardness ranges between 3.5 and 4, while calcite's is approximately 3. The density of dolomite also exceeds that of calcite, supporting its identification [136-138].

Quartz is another prominent mineral identified in the structural framework of the rock. It represents one of the most abundant polymorphic modifications of silicon dioxide in geological systems. Quartz crystallizes in the trigonal system and has a refractive index of 1.544. It is characterized by a Mohs hardness of 7 and a specific gravity of approximately 2.6 g/cm^3. In sedimentary environments, quartz is frequently intergrown with dolomitic minerals. Among its notable properties are its

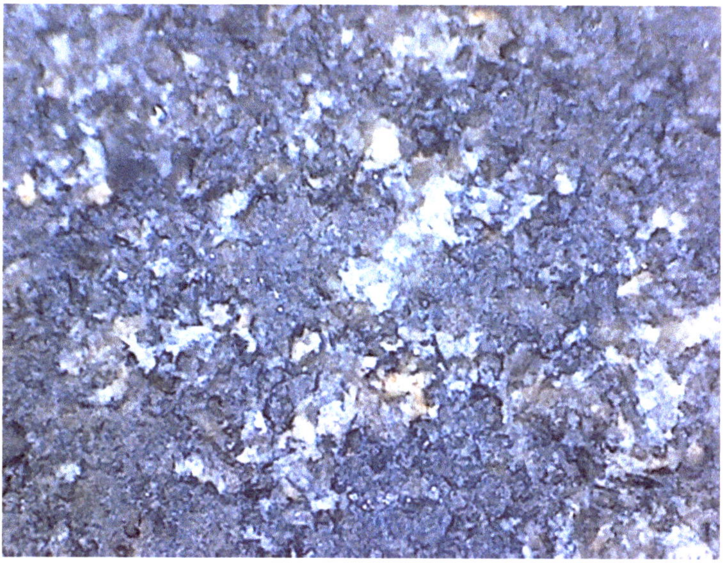

Fig. 3.11 Optical micrograph of low-grade phosphorites from the Akzhar deposit

Table 3.9 Chemical composition of the raw phosphate material from Karatau deposit

Component	P_2O_5	CaO	MgO	K_2O	Al_2O_3	Fe_2O_3	SiO_2
Content (%)	13.36	24.12	2.87	1.04	6.01	2.39	36.14

conchoidal fracture and the absence of true cleavage, which further facilitates its recognition in petrographic studies [139].

Among aluminosilicate minerals, muscovite was identified as a minor phase within the sample, occurring in the form of fine flakes. Muscovite belongs to the mica group and crystallizes in the monoclinic system. It is characterized by moderate elasticity and a refractive index of about 1.6. The crystal habit of muscovite is layered and platy, consistent with its basal cleavage and lamellar morphology. Its presence in low-grade phosphorites reflects secondary alteration and incorporation during diagenetic processes [140].

3.2.3 Low-Grade Phosphorites of the Karatau Deposit

The chemical composition of raw phosphate material is a critical parameter that determines its suitability for subsequent beneficiation and processing stages. Table 3.9 presents the elemental and oxide-based content of the sample selected for this study.

According to the data presented in Table 3.9, the relatively low content of phosphorus pentoxide ($P_2O_5 = 13.36\%$) indicates that the raw material does not meet the technological specifications established by the leading regional producer of phosphorus-based products in Kazakhstan and Central Asia—LLP "Kazphosphate" [141, 142]. As such, the material requires beneficiation before being considered for industrial processing.

The high content of calcium oxide ($CaO = 24.12\%$) suggests a significant presence of calcite ($CaCO_3$) or other carbonate minerals, which must be selectively removed or transformed during beneficiation. Furthermore, the presence of 6.01% Al_2O_3 points to the existence of aluminosilicate or clay minerals, while the high concentration of silicon dioxide ($SiO_2 = 36.14\%$) reveals the dominance of quartz or other silicate phases in the matrix.

To supplement the chemical analysis, scanning electron microscopy (SEM) combined with energy-dispersive X-ray spectroscopy (EDX) was employed to investigate the microstructure and elemental distribution of the raw phosphate sample. The results are detailed in Fig. 3.12 and Table 3.10, where the observed morphology and elemental spectra provide insights into the mineralogical heterogeneity and potential challenges for processing.

The SEM micrographs presented in Fig. 3.12, accompanied by EDX spectral analysis, illustrate the fine-grained microstructure characteristic of low-grade phosphorites from the Karatau basin. The EDX spectra indicate the presence of major elemental constituents including calcium (Ca), silicon (Si), aluminum (Al),

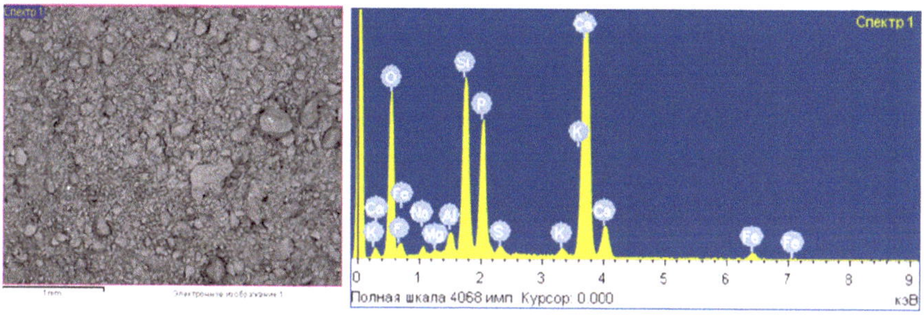

Fig. 3.12 SEM micrographs and EDS spectral analysis results of low-grade phosphorites from the Karatau deposit

Table 3.10 Elemental and oxide-recalculated composition of low-grade phosphorites from Karatau deposit

Element	Weight %	Recalculated oxide %
O	48.53	–
F	2.87	–
Na	0.78	1.05
Mg	0.23	0.38
Al	1.20	2.27
Si	11.59	26.51
P	5.61	13.02
S	0.52	1.30
K	0.56	0.67
Ca	26.57	34.37
Fe	1.56	2.23

phosphorus (P), iron (Fe), magnesium (Mg), potassium (K), titanium (Ti), and oxygen (O), suggesting the mineralogical heterogeneity of the sample.

The dominance of oxygen (48.53%) and calcium (26.57%) confirms the significant presence of carbonate minerals such as calcite. The pronounced silicon signal (11.59%) suggests a substantial amount of silicate phases, primarily quartz. The phosphorus signal (5.61%) confirms the presence of phosphate compounds, although its relatively lower intensity compared to calcium and silicon implies a limited concentration of phosphatic minerals. Other detected elements, including fluorine, sodium, magnesium, aluminum, sulfur, potassium, and iron, are present in minor proportions but nonetheless contribute to the complexity of the mineral matrix.

The data in Table 3.10 confirm that the primary compositional constituents of the sample are oxygen and calcium, consistent with a carbonate-rich lithology. The elevated silicon content indicates significant quartz or other silicate mineral inclusions. The phosphate content, although present, remains relatively low, supporting the classification of the material as low-grade. Minor components such as aluminum, magnesium, and iron oxides may stem from clay minerals, while traces of sodium, potassium, and sulfur suggest the presence of secondary or accessory phases.

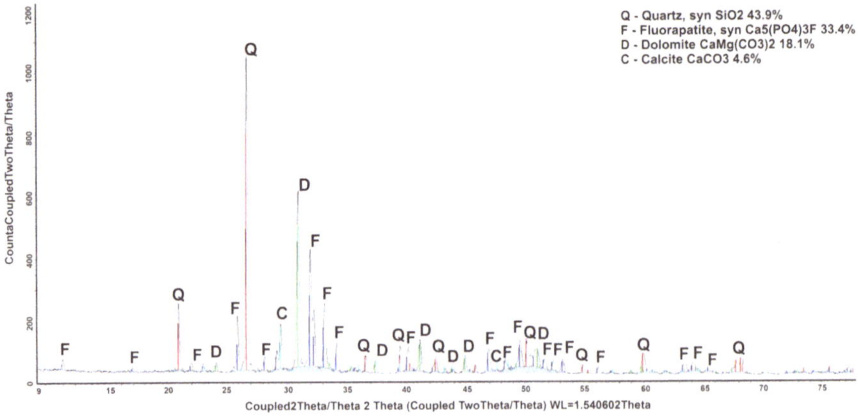

Fig. 3.13 X-ray diffraction pattern of low-grade phosphorite from the Karatau deposit

To obtain a more comprehensive picture of the sample's phase composition and crystalline structure, complementary X-ray diffraction (XRD) analysis was conducted, with results provided in Fig. 3.13.

The X-ray diffraction (XRD) analysis performed on the low-grade phosphorite sample from the Karatau Basin revealed the presence of four dominant crystalline phases: quartz (43.9%), fluorapatite (33.4%), dolomite (18.1%), and calcite (4.6%). These findings provide a detailed insight into the phase composition and are in strong agreement with earlier results obtained from chemical analysis and SEM-EDX spectra.

The high proportion of quartz observed via XRD corresponds with the substantial silicon content detected in both the elemental and oxide forms, confirming the presence of significant silicate phases. Fluorapatite is identified as the primary phosphate-bearing mineral, accounting for approximately one-third of the total mineral composition. Its presence substantiates the phosphorus levels measured during chemical and spectrometric evaluations.

Carbonate phases, specifically dolomite and calcite, are clearly indicated in the diffractogram and further support the interpretation of abundant calcium and magnesium oxides identified in previous SEM-EDX and wet chemical analyses. These minerals are known to significantly influence both the beneficiation process and the reactivity of the material during acid leaching.

The presence of these inorganic phases is also corroborated by Fourier-transform infrared (FTIR) spectroscopic data presented in Fig. 3.14, which confirm characteristic vibrational modes associated with phosphate, silicate, and carbonate structures. Together, these datasets provide a comprehensive understanding of the mineralogical characteristics of the raw phosphorite, underscoring its classification as low-grade due to the limited proportion of phosphorus-bearing phases.

The Fourier-transform infrared (FTIR) spectroscopy analysis of the low-grade phosphorite sample revealed distinct absorption bands corresponding to various

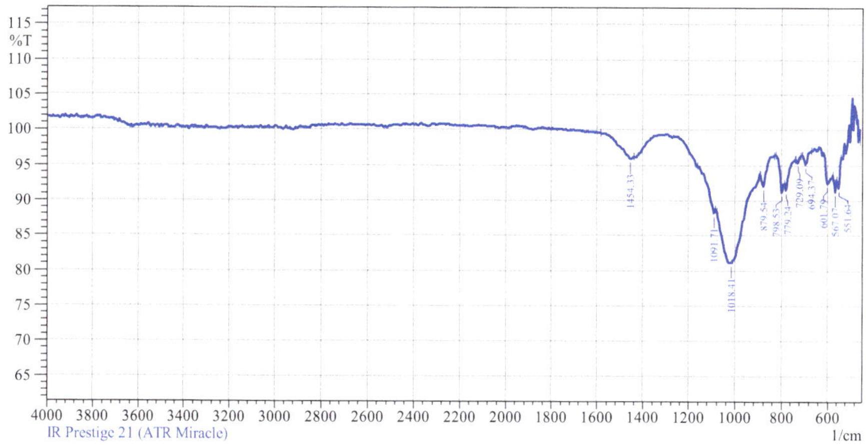

Fig. 3.14 FTIR spectra of low-grade phosphorite

functional groups associated with the major mineral components present in the material. These characteristic peaks provide detailed insights into the bonding environments and confirm the results of the chemical and XRD analyses.

The absorption peaks observed at 551.64 and 567.07 cm^{-1} are attributed to the bending vibrations of Si–O bonds typically associated with quartz structures. This observation aligns well with the high proportion of quartz detected in the X-ray diffraction data and further substantiates the presence of a significant silicate phase within the sample [133].

In the 694.37–729.09 cm^{-1} range, distinct bands correspond to the stretching vibrations of phosphate groups (PO_4^{3-}), which are indicative of fluorapatite minerals. These findings are consistent with previous XRD results that identified fluorapatite as a key phosphate-bearing mineral in the sample [133].

A prominent band at 1454.33 cm^{-1} is linked to the asymmetric stretching vibrations of carbonate groups (CO_3^{2-}), typically found in calcite and dolomite. This observation confirms the presence of carbonate phases detected during both chemical composition analysis (CaO and MgO contents) and XRD characterization [131].

Together, the FTIR data complement the XRD and SEM-EDX analyses, providing further confirmation of the complex mineralogical nature of the low-grade phosphorite. The overlapping signals of phosphate, silicate, and carbonate groups reflect the heterogeneous structure of the raw material and reinforce the necessity for selective beneficiation techniques to isolate valuable components effectively.

The subsequent section (Fig. 3.15) presents the thermogravimetric analysis results, which further elucidate the thermal behavior and phase transformations of the sample under controlled heating conditions.

The graph presented in Fig. 3.15 illustrates the results of the differential thermal analysis (DTA) and thermogravimetric analysis (TGA) performed on a sample of low-grade phosphorite. The blue curve tracks the mass loss of the sample as a function of increasing temperature. A significant drop in mass is observed within the

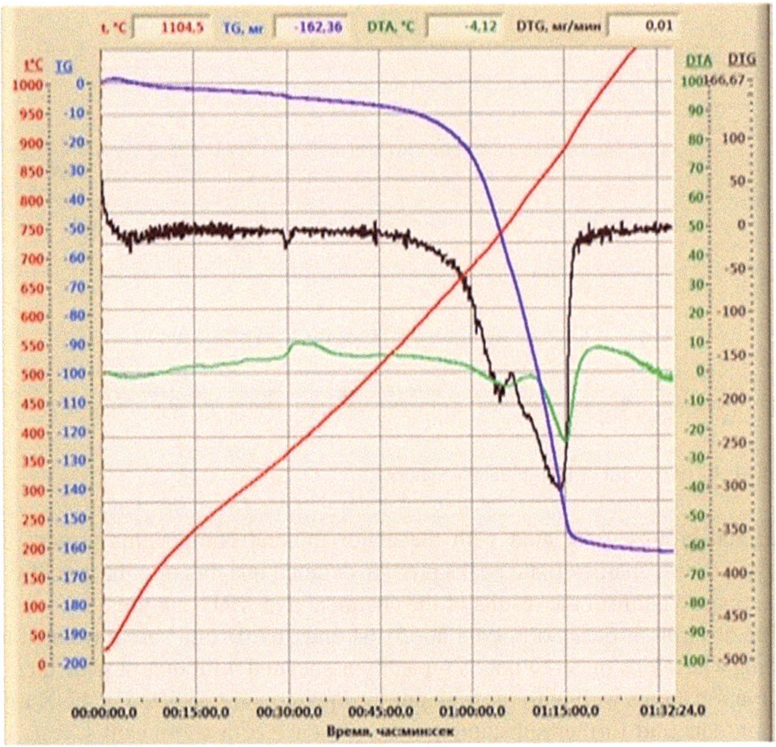

Fig. 3.15 Derivatogram of low-grade phosphorites from Karatau deposit

750–800 °C range. This pronounced mass loss is attributed to the decomposition of carbonate minerals such as calcite ($CaCO_3$) and dolomite ($CaMg(CO_3)_2$), a process which liberates carbon dioxide (CO_2). This phase transformation accounts for the reduction in mass. The green curve indicates thermal reactions within the sample—distinguishing between endothermic and exothermic events. A marked endothermic peak is evident around 750–800 °C, confirming the thermal decomposition of carbonate phases. This reaction absorbs heat, validating the interpretation of TG results. The purple curve represents the rate of mass loss with respect to temperature. The peak observed in the same 750–800 °C range signifies an accelerated loss of mass, indicative of intense decarbonation or dehydration processes taking place within the sample structure. The red curve outlines the programmed heating of the sample, reaching a maximum of approximately 1104.5 °C. This extended temperature range enables observation of transformations up to the thermal stability limits of the constituent mineral phases.

The simultaneous mass loss, endothermic signature, and differential thermogravimetry (DTG) peak within the 750–800 °C interval collectively provide strong evidence for the decomposition of carbonate phases—primarily dolomite and calcite. These thermal events are critical for understanding the stability and behavior of low-grade phosphorites during high-temperature processes such as roasting or acid leaching.

The thermal behavior described here is corroborated by mineralogical characterization shown in Fig. 3.16, which complements the interpretation by identifying the key phases involved in thermal transitions.

The microscopic images of low-grade phosphorite samples, obtained using the NEOPHOT-21 metallographic microscope at 450× magnification, reveal several key features that align with the findings from elemental analysis (EDX), X-ray diffraction (XRD), differential thermal analysis (DTA), and Fourier-transform infrared spectroscopy (FTIR). The images clearly demonstrate the presence of fine-grained particles dispersed throughout the matrix. This observation confirms the complex mineralogical composition of the sample, which is consistent with XRD results indicating significant proportions of quartz, fluorapatite, dolomite, and calcite. The granular morphology corresponds to the high content of silicate minerals, particularly quartz, identified as a major phase (43.9%). Both images show distinctly bright and dense regions, which likely represent heavier minerals. These zones are attributed to carbonate minerals such as calcite and dolomite, which were also identified in chemical and thermal analysis. The notable mass loss observed at 750–800 °C in the DTA is consistent with the thermal decomposition of these carbonate phases. The predominant dark tone in the matrix is indicative of the presence of clay minerals and phosphate phases. This is supported by FTIR spectra, where characteristic absorption bands associated with phosphate groups (PO_4^{3-} in fluorapatite) and carbonate groups (CO_3^{2-} in calcite and dolomite) were detected. The micrographs show clear heterogeneity in the distribution of mineral phases. This structural

Fig. 3.16 Optical micrograph of low-grade phosphorites from Karatau deposit

Table 3.11 Chemical composition of low-grade phosphorites from the Shilisai deposit

Component	Content, %
Total P_2O_5	18.12
CaO	31.19
MgO	1.72
K_2O	0.13
Al_2O_3	2.04
Fe_2O_3	0.97
SiO_2	23.10

non-uniformity is in full agreement with FTIR and XRD data, where a variety of functional groups and crystalline phases were detected. The presence of multiple mineralogical components—such as phosphate (fluorapatite), silicate (quartz), and carbonate (calcite and dolomite) minerals—underscores the complexity of the material and suggests potential challenges in beneficiation and processing [137].

These microscopic observations provide critical insight into the structural composition of low-grade phosphorites and support a comprehensive understanding of their mineralogical behavior during thermal and chemical treatment.

3.2.4 Low-Grade Phosphorites of the Shilisai Deposit

The Shilisai phosphate deposit, located in the western region of Kazakhstan within the Aktobe phosphorite basin, is characterized by low-grade phosphorite reserves. The chemical composition of representative samples collected from this deposit is presented in Table 3.11.

Based on the data in Table 3.11, the major components of the phosphorite sample include phosphorus pentoxide (P_2O_5) and calcium oxide (CaO), which are present in significant proportions of 18.12% and 31.19%, respectively. These values indicate the presence of apatite minerals and calcium-bearing phases. Magnesium oxide (MgO) accounts for 1.72%, while aluminum oxide (Al_2O_3) and ferric oxide (Fe_2O_3) are present in moderate amounts of 2.04% and 0.97%, respectively. The silica (SiO_2) content is relatively high at 23.10%, suggesting the presence of quartz or other silicate phases within the matrix.

Further mineralogical characterization was carried out through microstructural imaging under various magnification regimes. Elemental weight composition was also determined based on energy-dispersive spectroscopy (EDS) across four representative spectra. The results of these analyses are illustrated in Fig. 3.17 and summarized in Table 3.12.

The data derived from four representative spectra of Shilisai phosphorite samples reveal a mineral composition dominated by calcium, phosphorus, and silicon compounds. The phosphorus pentoxide content recalculated from elemental phosphorus averages 17.17%, affirming the economic viability of these phosphorites for fertilizer-grade processing.

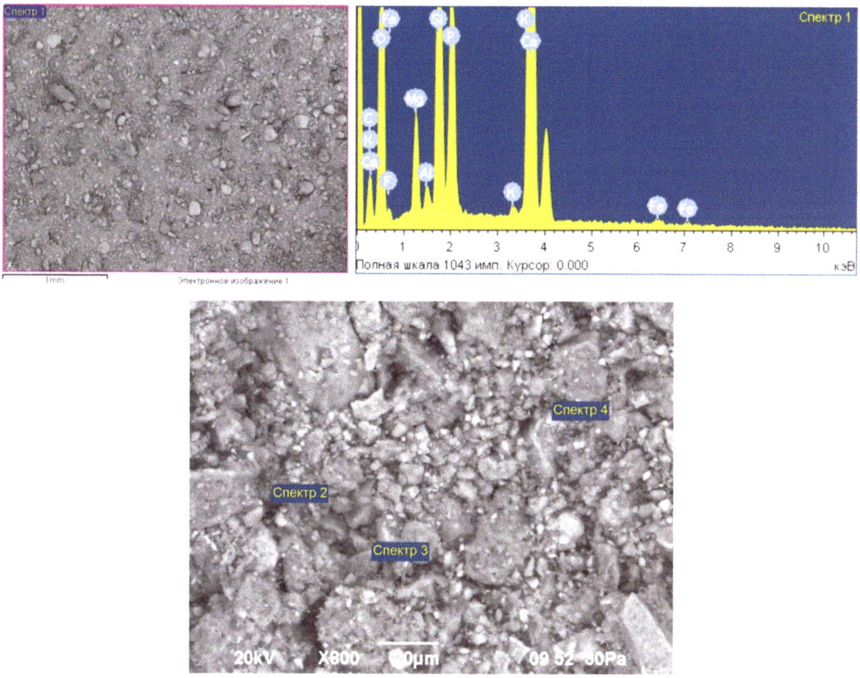

Fig. 3.17 SEM micrographs and EDS spectral analysis results of low-grade phosphorites from the Shilisai deposit

Calcium oxide concentration, derived from elemental Ca, is significant—33.08% on average—indicating the prevalence of apatite or calcite phases. Silicon content is also high, at 25.60%, suggesting the presence of quartz or other silicate structures.

Magnesium and aluminum oxides were measured at 2.61% and 1.87%, respectively, while iron oxide accounted for 0.61%. The presence of fluorine (1.56%) supports the identification of fluorapatite as a dominant phosphate-bearing phase.

Notably, carbon was present in selected spectra (2.34% average), which may be indicative of carbonate phases or organic matter traces. The results corroborate previous chemical data and support the mineralogical heterogeneity of the Shilisai deposit.

To further elucidate the compositional features of the Shilisai phosphorites, Fourier-transform infrared (FTIR) spectroscopy was performed using the IR Prestige system. The FTIR spectra provide structural insights into the molecular vibrations characteristic of phosphate, carbonate, and silicate functional groups. The detailed spectral data and interpretation are presented in Fig. 3.18 and Table 3.13.

The FTIR (Fourier-transform infrared) spectroscopic analysis of low-grade phosphorites from the Shilisai deposit was conducted to determine the vibrational characteristics and to identify the functional groups corresponding to various

Table 3.12 Elemental and oxide-recalculated composition of low-grade phosphorites from the Shilisai deposit

Element	Spectrum 1 (wt%)	Oxide (calc.%)	Spectrum 2	Oxide	Spectrum 3	Oxide	Spectrum 4	Oxide	Average wt%	Average oxide %
C	2.93	–	–	–	–	–	1.76	–	2.34	–
O	43.26	–	54.86	–	54.53	–	52.94	–	51.39	–
F	2.32	–	0.66	–	1.71	–	–	–	1.56	–
Mg	3.37	5.58	1.19	1.97	1.37	2.27	0.42	0.69	1.58	2.61
Al	1.60	3.02	0.87	1.64	0.11	0.20	1.38	2.60	0.99	1.87
Si	10.17	21.75	14.21	30.39	11.08	23.70	12.43	26.58	11.97	25.60
P	8.08	18.51	6.31	14.45	7.43	17.02	8.17	18.71	7.49	17.17
K	0.13	0.15	–	–	0.27	0.32	0.10	0.12	0.16	0.19
Ca	27.72	38.78	21.09	29.50	23.16	32.40	22.63	31.65	23.65	33.08
Fe	0.42	0.60	0.81	1.15	0.34	0.48	0.17	0.24	0.43	0.61

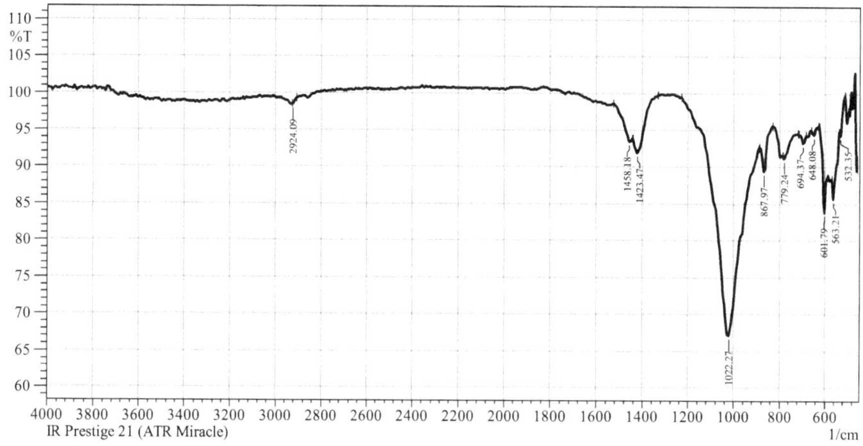

Fig. 3.18 FTIR spectra of low-grade phosphorites from the Shilisai deposit

Table 3.13 Peak positions of FTIR spectra

Peak	Intensity	Corr. intensity	Base (*H*)	Base (*L*)	Area	Corr. area
532.35	94.235	1.344	536.21	513.07	0.300	0.060
563.21	85.712	4.090	570.93	536.21	1.669	0.390
601.79	83.832	7.091	624.94	590.22	1.771	0.454
648.08	94.334	0.921	659.66	628.79	0.691	0.056
694.37	93.216	1.144	713.66	682.80	0.844	0.068
779.24	91.147	0.788	786.96	717.52	2.114	0.059
867.97	89.387	4.313	887.26	829.39	1.835	0.352
1022.27	67.203	28.033	1230.58	891.11	22.997	16.850
1423.47	91.889	2.984	1442.75	1334.74	1.885	0.490
1458.18	93.498	1.010	1527.62	1446.61	1.444	0.089
2924.09	98.374	0.260	2927.94	2908.65	0.104	0.009

mineral phases. The spectral interpretation was performed using IR Prestige equipment, and the results are summarized in Table 3.13.

The observed peaks in the 1000–1060 and 800–960 cm^{-1} regions with high intensity are indicative of Si–O–Si vibrations, confirming the presence of silicate compounds within the phosphate matrix. These bands are associated with the asymmetric stretching vibrations of silicon-oxygen tetrahedra commonly found in quartz and other silicate minerals.

A prominent peak around 550–570 cm^{-1} corresponds to fluorophosphate compounds, particularly those involving PO_4 groups bonded to fluorine ions, such as in fluorapatite [131–133].

Meanwhile, lower intensity bands observed in the 1435–1450 cm^{-1} range are characteristic of carbonate groups (CO_3^{2-}), often attributed to calcite or dolomite phases. These absorption features reflect the stretching modes of the C–O bonds in carbonate minerals and suggest their presence in minor amounts within the sample [134].

The identification of these key spectral features confirms the complex composition of the Shilisai phosphorites, which include silicates, fluorophosphates, and carbonates. This mineralogical heterogeneity is consistent with the sedimentary genesis of the deposit and is crucial for evaluating the appropriate beneficiation strategies.

To further determine the phase composition of the sample, complementary X-ray diffraction (XRD) analysis was performed. The results of the XRD investigation are presented in Fig. 3.19 and Table 3.14, allowing for cross-validation of the functional groups and crystalline phases inferred from FTIR data.

According to the results of X-ray diffraction analysis, the dominant crystalline phase in the low-grade phosphorite samples from the Shilisai deposit is carbonate-fluoroapatite, which accounts for 64.10% of the overall composition. This high proportion of carbonate-substituted apatite is indicative of phosphorites formed in biogenic environments, particularly algal-derived (gelatinous) phosphate sediments. Such phosphorites are characterized by partial isomorphic substitution of phosphate groups (PO_4^{3-}) with carbonate ions (CO_3^{2-}) in the apatite lattice structure [21].

The presence of 3.72% dolomite ($CaMg(CO_3)_2$) further confirms the carbonate-rich nature of the sample and validates the reliability and consistency of the mineralogical characterization. The identification of dolomite as a minor phase supports the conclusion that the examined phosphorite belongs to the class of carbonate-rich sedimentary phosphorites.

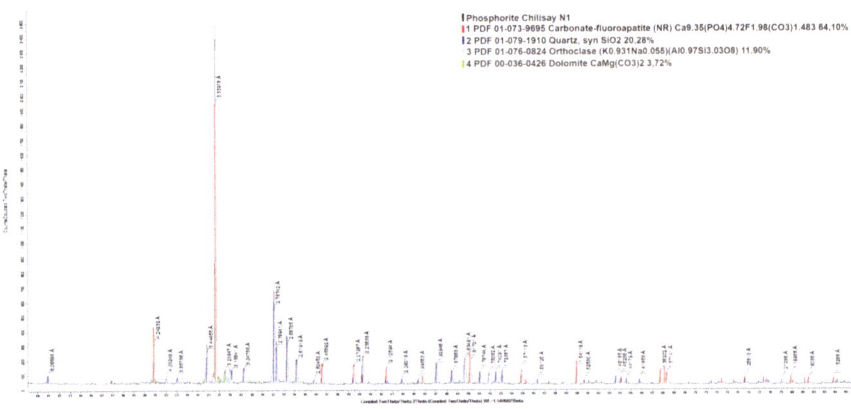

Fig. 3.19 X-ray diffraction (XRD) pattern of low-grade phosphorites from the Shilisai deposit

Table 3.14 Phases identified in Shilisai phosphorites based on XRD analysis

PDF reference	Identified phase	Content, %
PDF 01-073-9695	Carbonate-fluoroapatite (NR) $Ca_{9.35}(PO_4)_{4.72}F_{1.98}(CO_3)_{1.483}$	64.10
PDF 01-079-1910	Quartz, syn SiO_2	20.28
PDF 01-076-0824	Orthoclase $(K_{0.931}Na_{0.055})(Al_{0.97}Si_{3.03}O_8)$	11.90
PDF 00-036-0426	Dolomite, $CaMg(CO_3)_2$	3.72

In addition to apatite, quartz (SiO_2) is present in significant quantities—20.28%—indicating the sedimentary origin and silicate contamination in the phosphorite bed. Furthermore, orthoclase, a potassium feldspar from the alkali feldspar group, accounts for 11.90% of the sample. The presence of orthoclase further supports the classification of the material as a mixed silicate–phosphate sedimentary rock.

To complement the phase identification and determine thermal behavior, differential thermal analysis (DTA) was conducted on the sample. The thermograms obtained using the Q-1500 D derivatograph are presented in Fig. 3.20, which provides insights into dehydration, phase transformation, and decomposition events corresponding to the previously identified mineralogical constituents.

According to the results of differential thermal analysis (DTA) conducted at a maximum temperature of 2610 °C, the endothermic effects observed between 70 and 180 °C indicate the release of structurally bound water from the mineral matrix. This dehydration process reflects the thermal instability of hydrous phases present in the sample. The modification transformation of quartz was registered at around 200 °C and above, corresponding to the known α- to β-quartz phase transition. Starting from approximately 700 °C, the decomposition of carbonate-containing

Fig. 3.20 Derivatogram of low-grade phosphorites from the Shilisai deposit

compounds such as dolomite and calcite was initiated, with noticeable exothermic and endothermic transitions [133, 134].

Over the course of the 82-min thermal run, the total weight loss of the Shilisai phosphorite sample amounted to 6.67%, confirming the presence of volatile phases such as water and CO_2 and providing insight into the decomposition behavior of the constituent mineral phases.

In the course of mineralogical characterization, phosphate minerals were found to occur in nodular (concretionary) forms. A substantial portion of the sample was composed of quartz, consistent with sedimentary origin and indicating the presence of silicate-rich matrices. Additionally, glauconite—a representative of the hydromica group—was identified among the mineral constituents. Glauconite is an unstable, heterogeneous iron aluminosilicate mineral comprising quartz and potassium oxides. It is commonly encountered in marine sedimentary environments and is known for its soft texture (Mohs hardness: 2–3) and a typical density of approximately 2.6 g/cm^3 [137–139].

Dolomite was also identified in trace quantities. Due to morphological similarities, it may be mistaken for calcite under microscopic analysis. However, its distinguishing characteristic lies in its ordered crystal structure, which results from the alternate arrangement of calcium and magnesium cations along the crystallographic *c*-axis [140].

The optical mineralogical microstructure of the Shilisai phosphorite sample is illustrated in Fig. 3.21, demonstrating the nodular phosphate textures and the surrounding quartz and dolomite matrices.

The microscopic image in Fig. 3.21 represents the mineralogical structure of low-grade phosphorite samples from the Shilisai deposit, obtained using the NEOPHOT-21 metallographic microscope at 450× magnification. The image

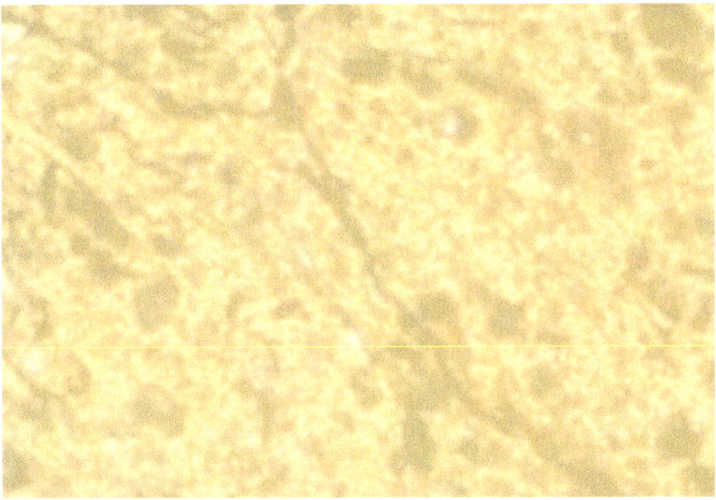

Fig. 3.21 Optical micrograph of low-grade phosphorites from the Shilisai deposit

distinctly shows the spatial distribution and textural features of the key mineral phases present in the sample.

Several primary mineral groups were identified and labeled: phosphorite nodules are observed as the dominant phase in the sample. These minerals are typically rounded or sub-rounded in shape and exhibit a lighter optical contrast, indicating the presence of carbonate-fluoroapatite ($Ca_{9.35}(PO_4)_{4.72}F_{1.98}(CO_3)_{1.483}$), which was previously confirmed by XRD analysis (see Table 3.14). These nodules are tightly aggregated, indicating recrystallization and partial diagenetic alteration. Dolomite grains are observed in interstitial zones and appear in darker, more compact forms. Their presence correlates with the thermally induced decomposition signatures of carbonate phases recorded in the DTA curves (see Fig. 3.20). Dolomite ($CaMg(CO_3)_2$) displays a fine to medium crystal size and contributes to the overall porosity of the sample. Quartz crystals are identified by their angular morphology and relatively high brightness under polarized light. Quartz is distributed throughout the matrix and accounts for approximately 20.28% of the total composition, supporting the findings of both elemental and diffractometric studies. Glauconite minerals, a representative of the hydromica group, are visible as fine, greenish-gray platy structures. These occur in minor amounts and are commonly associated with iron-aluminosilicate phases. Glauconite's occurrence suggests sedimentary conditions conducive to slow accumulation and indicates a marine depositional environment.

This mineralogical image supports previous analytical results and further verifies the heterogeneous composition of the Shilisai phosphorites, characterized by significant textural and structural complexity.

Chapter 4
Results and Discussion

4.1 Chemistry and Thermodynamics of the Beneficiation of Low-Grade Phosphorites with Organic Acids

The chemical interaction between phosphate raw materials and acetic acid involves a series of exchange reactions with the major mineralogical components present in the ore. These reactions can be expressed through the following chemical equations (4.1)–(4.7):

$$Ca_5(PO_4)_3F + 10CH_3COOH = 5Ca(CH_3COO)_2 + 3H_3PO_4 + HF \quad (4.1)$$

$$CaCO_3 + 2CH_3COOH = Ca(CH_3COO)_2 + H_2O + CO_2 \quad (4.2)$$

$$MgCO_3 + 2CH_3COOH = Mg(CH_3COO)_2 + H_2O + CO_2 \quad (4.3)$$

$$K_2O + 2CH_3COOH = 2KCH_3COO + H_2O \quad (4.4)$$

$$Al_2O_3 + 6CH_3COOH = 2Al(CH_3COO)_3 + 3H_2O \quad (4.5)$$

$$Fe_2O_3 + 6CH_3COOH = 2Fe(CH_3COO)_3 + 3H_2O \quad (4.6)$$

$$SiO_2 + 4CH_3COOH = Si(CH_3COO)_4 + 2H_2O \quad (4.7)$$

The reaction between fluorapatite and acetic acid (Eq. 4.1) results in the formation of soluble calcium acetate and phosphoric acid, alongside the evolution of hydrofluoric acid in the gas phase. Calcium and magnesium carbonates present in the ore undergo acid decomposition with the release of carbon dioxide, leading to the formation of their respective acetate salts, water, and gas.

Potassium and aluminum-containing compounds are typically present in the mineral phase as nepheline-like structures, and thus participate in acetic acid reactions yielding potassium and aluminum acetates. The reaction pathways outlined

Y. Raiymbekov, *Enrichment of Low-Grade Phosphorites with Organic Acids*, https://doi.org/10.1007/978-3-032-02910-2_4

above are based on established chemical principles and data from relevant literature sources. Consequently, a reaction mechanism of acetic acid interaction with phosphate-bearing minerals has been proposed (Fig. 4.1).

According to this mechanism, the selective leaching of the ore proceeds primarily through the dissolution of carbonate, potassium, and alumina phases, leaving behind a residue enriched in phosphate phases ($Ca_5(PO_4)_3F$), silica (SiO_2), and less reactive compounds such as CaF_2 and Fe_2O_3. The remaining solid fraction represents the insoluble portion, while the dissolved portion includes the acetates of the aforementioned elements.

Acetic acid, being a weak monoprotic carboxylic acid ($pK_a = 1.8 \times 10^{-5}$), acts as a polar protic solvent capable of donating protons (H^+) and facilitating ion exchange and dissolution processes. This behavior is consistent with other findings reported in the literature [114–119], which support the validity of these reaction pathways and their thermodynamic feasibility.

The subsequent section presents a thermodynamic evaluation of the above-described reactions and assesses their spontaneity, enthalpy changes, and equilibrium parameters under process-relevant conditions.

To perform the thermodynamic analysis of the enrichment process involving the reaction of phosphate raw materials and their associated compounds with acetic acid, the HSC 9.3 software suite was employed. This software system contains a comprehensive database of over 17,000 chemical compounds and operates based on the principles of maximum entropy and Gibbs free energy minimization.

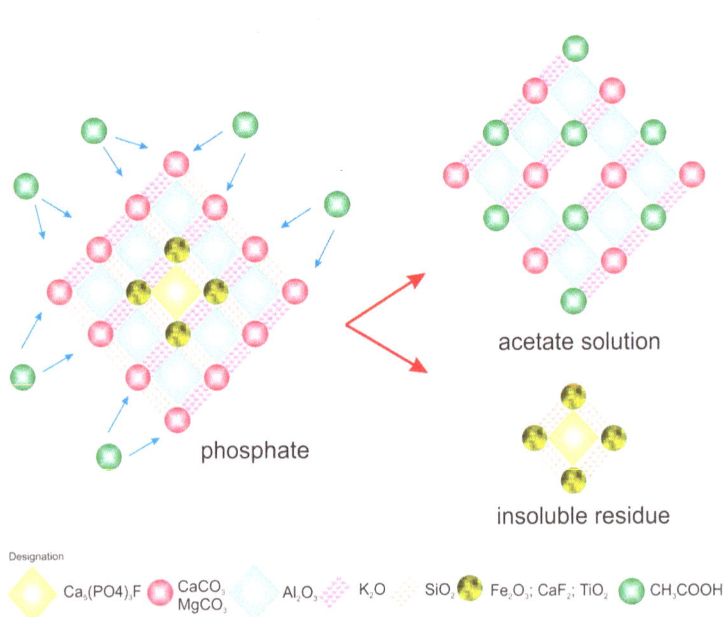

Fig. 4.1 Mechanism of reactions (4.1)–(4.7)

The thermodynamic evaluation was conducted within the temperature range of 293–333 K, which corresponds to typical processing conditions. This analysis focused on the viability of chemical reactions occurring during the acetic acid-based beneficiation of low-grade phosphorites and processing residues. The results of the thermodynamic simulations for reactions between acetic acid and individual components of the phosphate raw material are presented in Figs. 4.2, 4.3, 4.4, 4.5, 4.6 and 4.7 and Tables 4.1, 4.2, 4.3, 4.4, 4.5 and 4.6.

The data presented in Fig. 4.2 and Table 4.1 indicate that the reaction of fluorapatite with acetic acid is thermodynamically unfavorable across the entire temperature range examined. This conclusion is substantiated by the consistently positive values of the Gibbs free energy change ($\Delta G°$), which increases from 357.681 kJ/mol at 293 K to 401.544 kJ/mol at 333 K. Such high and increasing $\Delta G°$ values confirm that the reaction does not proceed spontaneously under the given conditions.

Interestingly, at approximately 313 K, a noticeable discontinuity in both the enthalpy ($\Delta H°$) and entropy ($\Delta S°$) values is observed. This anomaly likely results from entropic contributions associated with the breakdown of complex molecular structures and the liberation of gaseous or ionic species, which aligns with the physical interpretation of entropy as a measure of molecular disorder and dispersal.

In contrast, the reactions involving the dissolution of calcium and magnesium carbonates contained in the phosphate raw material exhibit negative values of $\Delta G°$ across the investigated temperature range, as shown in Fig. 4.3 and Table 4.2. These negative Gibbs free energy changes provide thermodynamic evidence that reactions (4.2) and (4.3) are feasible and spontaneous under the specified conditions. The observed spontaneity of these secondary reactions supports the selective leaching mechanism, where carbonates are more readily dissolved than the phosphate phase.

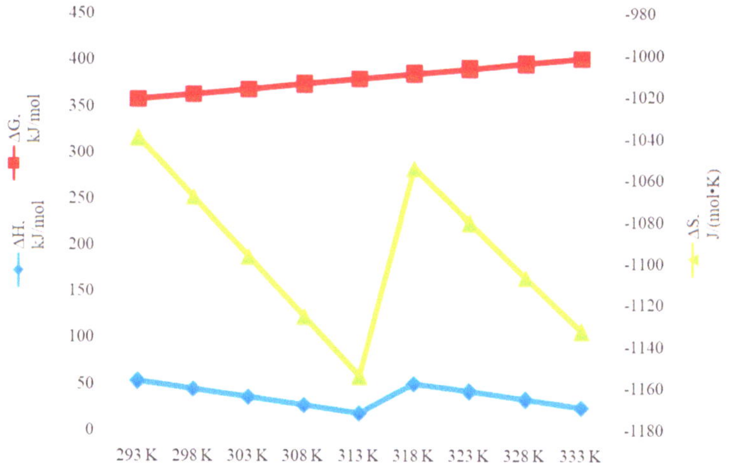

Fig. 4.2 Effect of temperature on the thermodynamic parameters of the reaction (4.1) between phosphate raw material and acetic acid

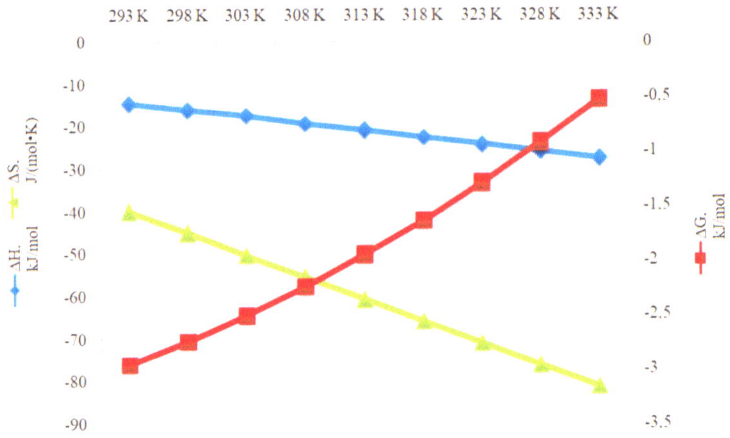

Fig. 4.3 Effect of temperature on the thermodynamic parameters of the reaction (4.2) between calcium carbonate and acetic acid

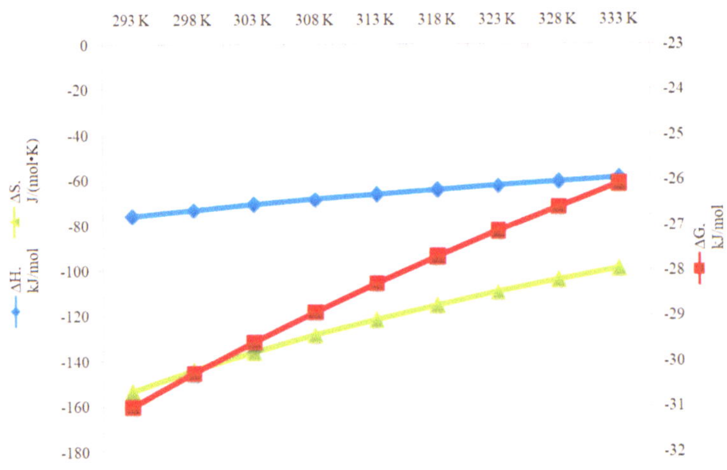

Fig. 4.4 Effect of temperature on the thermodynamic parameters of the reaction (4.3) between magnesium carbonate and acetic acid

The thermodynamic analysis of reaction (4.2), involving the interaction of calcium carbonate with acetic acid, reveals that the process is favorable and spontaneous over the entire studied temperature range (293–333 K). The Gibbs free energy change ($\Delta G°$) remains negative throughout, decreasing from −2.960 kJ/mol at 293 K to −0.533 kJ/mol at 333 K. Although $\Delta G°$ becomes less negative with rising temperature, its negativity still confirms the spontaneity of the process.

The enthalpy values ($\Delta H°$) decrease with increasing temperature, reflecting the exothermic nature of the reaction. Simultaneously, the entropy change ($\Delta S°$) becomes more negative, indicating a decrease in system randomness, likely due to

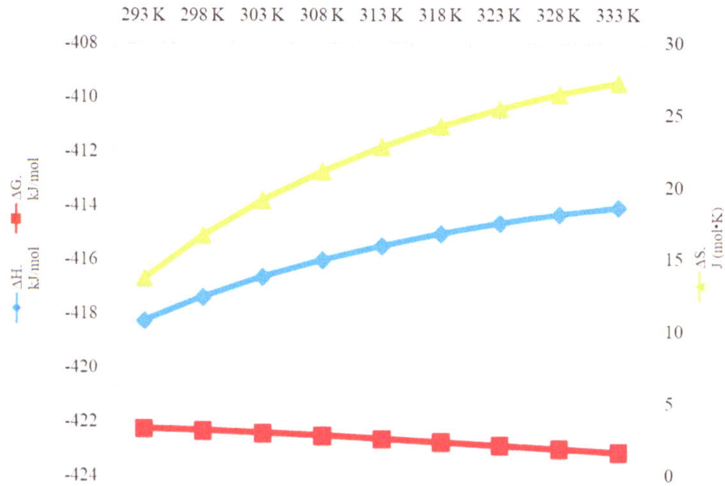

Fig. 4.5 Effect of temperature on the thermodynamic parameters of the reaction (4.4) between potassium oxide and acetic acid

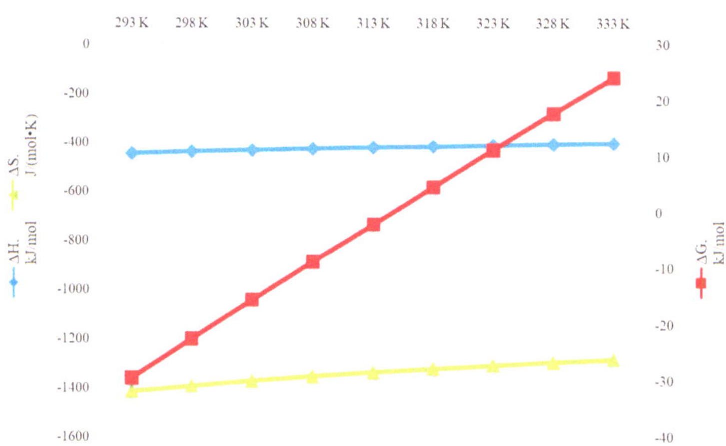

Fig. 4.6 Effect of temperature on the thermodynamic parameters of the reaction (4.5) between aluminum oxide and acetic acid

the formation of structured aqueous products such as calcium acetate and water, despite the evolution of gaseous CO_2.

The equilibrium constant (K) and its logarithm (log K) decline with temperature, suggesting a reduced reaction extent at higher temperatures. However, the magnitude of K remains >1 across all temperatures, confirming that the forward reaction remains thermodynamically dominant.

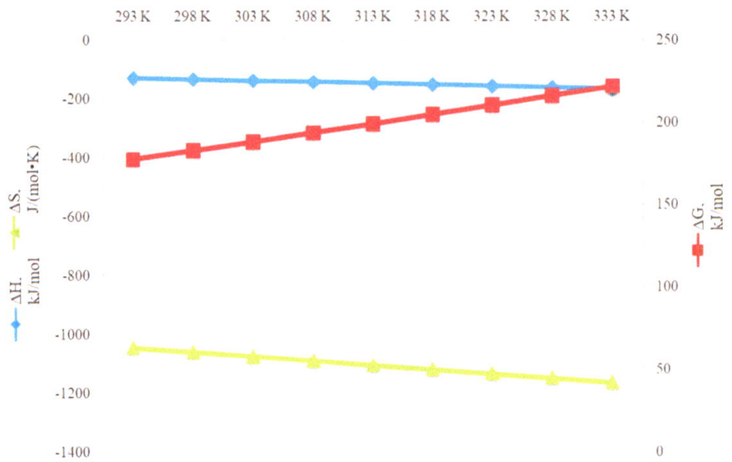

Fig. 4.7 Effect of temperature on the thermodynamic parameters of the reaction (4.6) between iron(III) oxide and acetic acid

Table 4.1 Thermodynamic parameters of reaction (4.1)

T, K	Δ*H*°, kJ/mol	Δ*S*°, J/(mol·K)	Δ*G*°, kJ/mol	*K*	log(*K*)
293	53.133	−1039.413	357.681	1.695E−064	−63.771
298	44.652	−1068.112	362.950	2.374E−064	−63.625
303	36.013	−1096.863	368.362	3.106E−064	−63.508
308	27.243	−1125.568	373.918	3.809E−064	−63.419
313	18.367	−1154.154	379.618	4.392E−064	−63.357
318	49.862	−1054.329	385.139	5.394E−064	−63.268
323	41.413	−1080.692	390.476	7.048E−064	−63.152
328	32.902	−1106.838	395.945	8.704E−064	−63.060
333	24.340	−1132.745	401.544	1.019E−064	−62.992

Table 4.2 Thermodynamic parameters for reaction (4.2)

T, K	Δ*H*°, kJ/mol	Δ*S*°, J/(mol·K)	Δ*G*°, kJ/mol	*K*	log(*K*)
293	−14.670	−39.967	−2.960	3.370E+000	0.528
298	−16.191	−45.113	−2.747	3.013E+000	0.482
303	−17.478	−50.295	−2.508	2.707E+000	0.432
308	−19.335	−55.490	−2.244	2.402E+000	0.381
313	−20.948	−60.684	−1.953	2.119E+000	0.326
318	−22.582	−65.865	−1.637	1.858E+000	0.269
323	−24.236	−71.024	−1.295	1.620E+000	0.209
328	−25.905	−76.154	−0.927	1.405E+000	0.148
333	−27.590	−81.251	−0.533	1.212E+000	0.084

Table 4.3 Thermodynamic parameters for reaction (4.3)

T, K	$\Delta H°$, kJ/mol	$\Delta S°$, J/(mol·K)	$\Delta G°$, kJ/mol	K	log(K)
293	−75.949	−153.386	−31.007	3.375E+005	5.528
298	−73.193	−144.054	−30.264	2.020E+005	5.305
303	−70.704	−135.772	−29.565	1.251E+005	5.097
308	−68.426	−128.313	−28.905	7.990E+004	4.903
313	−66.313	−121.508	−28.281	5.248E+004	4.720
318	−64.336	−115.240	−27.689	3.537E+004	4.549
323	−62.475	−109.433	−27.128	2.440E+004	4.387
328	−60.722	−104.049	−26.594	1.720E+004	4.236
333	−59.082	−99.084	−26.087	1.237E+004	4.092

Table 4.4 Thermodynamic parameters for reaction (4.4)

T, K	$\Delta H°$, kJ/mol	$\Delta S°$, J/(mol·K)	$\Delta G°$, kJ/mol	K	log(K)
293	−418.250	13.708	−422.266	1.932E+075	75.286
298	−417.370	16.686	−422.343	1.087E+074	74.036
303	−416.637	19.126	−422.432	6.758E+072	72.830
308	−416.020	21.146	−422.533	4.620E+071	71.665
313	−415.495	22.837	−422.643	3.453E+070	70.538
318	−415.046	24.262	−422.761	2.809E+069	69.448
323	−414.661	25.462	−422.886	2.475E+068	68.394
328	−414.337	26.460	−423.015	2.353E+067	67.372
333	−414.072	27.260	−423.150	2.405E+066	66.381

Table 4.5 Thermodynamic parameters for reaction (4.5)

T, K	$\Delta H°$, kJ/mol	$\Delta S°$, J/(mol·K)	$\Delta G°$, kJ/mol	K	log(K)
293	−443.821	−1413.930	−29.540	1.84E+005	5.267
298	−437.023	−1390.918	−22.529	8.898E+003	3.494
303	−430.945	−1370.689	−15.626	4.943E+002	2.694
308	−425.430	−1352.635	−8.818	3.131E+001	1.496
313	−420.361	−1336.309	−2.097	2.238E+000	0.350
318	−415.655	−1321.391	4.547	1.791E−001	−0.747
323	−411.261	−1307.679	11.119	1.591E−002	−1.798
328	−407.158	−1295.072	17.626	1.559E−003	−2.807
333	−403.353	−1283.557	24.072	1.674E−004	−3.776

Table 4.6 Thermodynamic values for reaction (4.6)

T, K	$\Delta H°$, kJ/mol	$\Delta S°$, J/(mol·K)	$\Delta G°$, kJ/mol	K	log(K)
293	−129.129	−1046.666	177.544	2.216E–032	−31.654
298	−133.517	−1061.517	182.815	8.971E–033	−32.047
303	−138.010	−1076.467	188.160	3.632E–033	−32.440
308	−142.589	−1091.456	193.580	1.471E–033	−32.832
313	−147.241	−1106.438	199.074	5.956E–034	−33.225
318	−151.954	−1121.378	204.644	2.412E–034	−33.618
323	−156.721	−1136.250	210.288	9.773E–035	−34.010
328	−161.533	−1151.035	216.006	3.960E–035	−34.402
333	−166.387	−1165.720	221.798	1.606E–035	−34.794

This behavior indicates that acetic acid effectively dissolves calcium carbonate within phosphate raw materials, facilitating selective leaching, and supports the feasibility of using acetic acid in beneficiation processes targeting the removal of carbonate phases.

The thermodynamic analysis of the interaction between magnesium carbonate and acetic acid reveals a consistently favorable process over the temperature range of 293–333 K. The Gibbs free energy change ($\Delta G°$) remains negative at all temperatures, confirming the spontaneity of reaction (4.3). At 293 K, $\Delta G°$ is −31.007 kJ/mol, and this value slightly increases to −26.087 kJ/mol at 333 K. Despite the moderate increase in $\Delta G°$ with temperature, the reaction remains thermodynamically favorable across the studied interval.

Both the enthalpy ($\Delta H°$) and entropy ($\Delta S°$) changes indicate exothermic and entropically favorable behavior, though with decreasing absolute values at higher temperatures. Compared to calcium carbonate, magnesium carbonate shows more negative $\Delta G°$ values throughout, indicating a greater thermodynamic tendency for leaching under identical conditions.

The equilibrium constant (K) for this reaction is significantly greater than 1 at all temperatures, with values exceeding 10^4 even at higher temperatures, thus reaffirming the strong thermodynamic drive of the reaction. The decline in K and log(K) values with rising temperature reflects the slightly reduced favorability but does not compromise the reaction's viability.

Overall, these results suggest that magnesium carbonate within phosphate raw materials reacts more readily with acetic acid than calcium carbonate. This supports a more efficient selective leaching mechanism for magnesium components during beneficiation. Additional analyses on the thermodynamic behavior of potassium- and aluminum-containing compounds in phosphate ore under similar conditions are presented in Figs. 4.5 and 4.6 and Tables 4.4 and 4.5.

The thermodynamic evaluation of the reaction between potassium oxide (K_2O) and acetic acid (CH_3COOH) presented in Table 4.4 and Fig. 4.5 demonstrates that the process is highly favorable over the entire temperature range of 293–333 K. The

Gibbs free energy (ΔG°) remains extremely negative throughout, with a value of −422.266 kJ/mol at 293 K and decreasing slightly to −423.150 kJ/mol at 333 K. These negative values strongly confirm the spontaneous nature of the reaction, regardless of the increase in temperature.

The enthalpy change (ΔH°) remains constant in its exothermic trend, showing minimal variation (from −418.250 to −414.072 kJ/mol), which suggests that the process is strongly driven by enthalpic factors. Concurrently, entropy (ΔS°) gradually increases with temperature, which supports enhanced disorder due to the generation of water molecules and soluble acetate salts.

The equilibrium constants (K) exhibit extremely high values at all temperatures, ranging from 1.932E+075 at 293 K to 2.405E+066 at 333 K. While these values show a decreasing trend as temperature increases, they remain orders of magnitude above unity, indicating overwhelming favorability for product formation.

From a mechanistic standpoint, the significant negative Gibbs free energy implies not only high spontaneity but also a negligible activation barrier under standard conditions. The increase in entropy indicates that acetic acid promotes disruption of the oxide structure and facilitates salt formation.

In summary, the interaction between potassium oxide and acetic acid represents a thermodynamically favorable process that proceeds efficiently across the entire temperature range considered. These data support the practical feasibility of potassium extraction from phosphate ores containing K_2O via mild organic acid leaching. Further consideration of thermodynamic behavior in systems involving aluminum oxide is discussed in Fig. 4.6 and Table 4.5.

The thermodynamic data presented in Table 4.5 and illustrated in Fig. 4.6 indicate that both potassium and aluminum oxides are capable of undergoing chemical interaction with acetic acid, though to markedly different extents. While potassium oxide exhibits strong and sustained reactivity across the entire temperature interval with significantly negative ΔG° values (see Table 4.4), aluminum oxide shows a steep thermodynamic limitation upon temperature increase.

At lower temperatures (293–313 K), the reaction between Al_2O_3 and CH_3COOH remains thermodynamically favorable, with Gibbs free energy (ΔG°) ranging from −29.540 to −2.097 kJ/mol. However, at 318 K and above, the ΔG° value becomes positive (e.g., +4.547 kJ/mol at 318 K), indicating the onset of non-spontaneity. This transition suggests a strong temperature dependence driven by entropy suppression, as evidenced by the high negative ΔS° values, which decrease in magnitude with rising temperature.

The sharp decline in equilibrium constant (K) with temperature—from 1.84E+005 at 293 K to 1.674E–004 at 333 K—further reinforces the inverse correlation between temperature and reaction feasibility. These observations imply that while acetic acid can react with aluminum oxide under controlled thermal conditions, its efficiency is significantly diminished beyond 313 K.

The reactivity of aluminum oxide in acetic acid solutions may be rationalized by the presence of aluminum-bearing minerals such as nepheline [$(Na,K)AlSiO_4$] in phosphate ore matrices [140]. Nepheline, a framework aluminosilicate mineral, is known to undergo partial dissolution in weak organic acids, including acetic acid.

This phenomenon has been previously documented by Soviet researchers such as Dorfman et al. [143], who confirmed nepheline's solubility in weak acids under specific thermodynamic conditions. Their findings support the feasibility of partial leaching of aluminum oxides from phosphate ores using acetic acid under optimized low-temperature regimes.

In contrast, subsequent analysis (as shown in Fig. 4.7 and Table 4.6) reveals that iron(III) oxide (Fe_2O_3), another common impurity in phosphate raw materials, does not engage in spontaneous interaction with acetic acid in the tested temperature interval. The $\Delta G°$ values remain consistently positive, indicating the reaction is thermodynamically unfavorable under the studied conditions.

The interaction between iron(III) oxide (Fe_2O_3) and acetic acid was investigated over a broad temperature range (293–333 K) in order to assess its thermodynamic feasibility. The calculated Gibbs free energy values ($\Delta G°$), as outlined in Table 4.6, are unequivocally positive throughout the examined temperature interval, ranging from +177.544 kJ/mol at 293 K to +221.798 kJ/mol at 333 K. This consistent trend confirms that the reaction is highly nonspontaneous under standard conditions and becomes increasingly unfavorable as temperature rises.

Despite the strongly exothermic enthalpy changes ($\Delta H° < 0$), the large negative entropy contributions ($\Delta S° < -1000$ J/mol·K) dominate the thermodynamic behavior of the system. The positive $\Delta G°$ values indicate that the disorder introduced by the reaction does not compensate for the enthalpic benefit, likely due to the low solubility and structural rigidity of Fe_2O_3 in acetic acid environments.

Additionally, the equilibrium constant (K) diminishes dramatically with increasing temperature—from 2.216E–32 at 293 K to just 1.606E–35 at 333 K—corroborating the conclusion that no significant progress of the reaction occurs under practical thermal conditions. The logarithmic values of K ($\log(K)$) show a linear decline, indicating a thermodynamically consistent suppression of reactivity.

From an applied standpoint, these findings are crucial for designing selective leaching protocols for phosphate raw materials. Iron oxides, being non-reactive under weak acid treatment, are expected to remain in the solid residue during acetic acid-based beneficiation. This inertness can be strategically exploited to concentrate more reactive mineral components (e.g., calcium, magnesium, potassium compounds) while minimizing iron contamination in the leachate.

In summary, the thermodynamic analysis decisively rules out the spontaneous dissolution of Fe_2O_3 in acetic acid within the temperature interval studied. Its high $\Delta G°$ values and vanishingly small equilibrium constants make it a thermodynamically stable phase under mild acid conditions. This behavior is consistent with literature findings and underscores the necessity of stronger complexing agents or higher energy inputs (e.g., hydrothermal conditions) for the mobilization of ferric iron in heterogeneous mineral matrices.

The thermodynamic data summarized in Table 4.7 and depicted in Fig. 4.8 demonstrate that the reaction between silicon dioxide (SiO_2) and acetic acid (CH_3COOH) is highly unfavorable under the given temperature range. This conclusion is strongly supported by the positive values of the Gibbs free energy ($\Delta G°$), which remain exceedingly high across all measured temperatures—from +848.17 kJ/mol at 293 K to +834.43 kJ/mol at 333 K.

The reactivity of silicon dioxide with weak organic acids such as acetic acid is extremely limited due to the inert nature of the Si–O bonds in the crystal lattice. Silicon dioxide typically resists acid attack except under conditions involving strong mineral acids like sulfuric, hydrochloric, or nitric acid, and even then, requires elevated temperatures or pressure to achieve notable dissolution. As a result, the transformation of SiO_2 under ambient conditions is thermodynamically implausible.

This finding aligns with known chemical behavior, where silicon-containing minerals require aggressive chemical environments to undergo significant reactivity. The poor solubility and low reactivity of SiO_2 in acetic acid further indicate that any interaction would be minimal and likely confined to the surface or amorphous phases, if any.

Considering all analyzed components of phosphate rock and their interaction with acetic acid, the following conclusions can be drawn regarding thermodynamic feasibility:

Table 4.7 Thermodynamic parameters of reaction (4.7)

T, K	$\Delta H°$, kJ/mol	$\Delta S°$, J/(mol·K)	$\Delta G°$, kJ/mol	K	log(K)
293	948.758	343.293	848.173	6.011E–152	−151.221
298	948.803	343.446	846.456	4.141E–149	−148.383
303	948.831	343.537	844.739	2.300E–146	−145.638
308	948.843	343.577	843.021	1.041E–143	−142.983
313	948.842	343.575	841.303	3.874E–141	−140.412
318	948.830	343.538	839.585	1.197E–138	−137.922
323	948.809	343.472	837.868	3.097E–136	−135.509
328	948.780	343.381	836.151	6.762E–134	−133.170
333	948.743	343.270	834.434	1.255E–131	−130.901

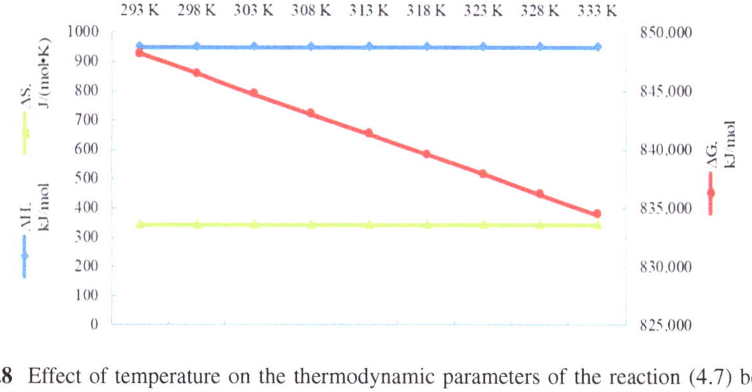

Fig. 4.8 Effect of temperature on the thermodynamic parameters of the reaction (4.7) between silicon dioxide and acetic acid

1. Reactions with Positive $\Delta G°$ (Not Thermodynamically Favorable):
   (a) Fluorapatite and iron(III) oxide (Fe_2O_3) display positive $\Delta G°$ values across the studied temperature range, signifying that their reactions with acetic acid are not spontaneous under standard conditions.
2. Reactions with Negative $\Delta G°$ (Thermodynamically Favorable):
   (a) Calcium carbonate ($CaCO_3$), magnesium carbonate ($MgCO_3$), potassium oxide (K_2O), and aluminum oxide (Al_2O_3) all show negative $\Delta G°$ values at lower temperatures, confirming their spontaneous reactivity with acetic acid. The thermodynamic favorability follows the order:
   $K_2O > MgCO_3 > CaCO_3 > Al_2O_3$.
   (b) These reactions are also exothermic, as indicated by their negative $\Delta H°$ values, highlighting the release of heat during the process.

In summary, the thermodynamic evaluation provides valuable insights into the selective reactivity of phosphate rock constituents with acetic acid. This data can inform further experimental optimization and potential industrial application in phosphate ore beneficiation and organic acid leaching processes.

The process of selective leaching is based on the targeted dissolution of carbonate-containing minerals present in phosphate raw materials by means of weak organic acids. This approach leverages the acid-base reactivity of calcium carbonate ($CaCO_3$) and magnesium carbonate ($MgCO_3$) with organic acids such as citric acid ($C_6H_8O_7$) and formic acid (HCOOH). During the reaction, soluble salts—specifically calcium and magnesium citrates or formates—are formed, accompanied by the release of carbon dioxide (CO_2) and water (H_2O) as byproducts.

The following reactions exemplify this behavior (4.8)–(4.11):

$$3CaCO_3 + 2H_3C_6H_5O_7 \rightarrow Ca_3\left(C_6H_5O_7\right)_2 + 3H_2O + 3CO_2\uparrow\left(\Delta G = -56\,kJ/mol\right) \tag{4.8}$$

$$3MgCO_3 + 2H_3C_6H_5O_7 \rightarrow Mg_3\left(C_6H_5O_7\right)_2 + 3H_2O + 3CO_2\uparrow\left(\Delta G = -117\,kJ/mol\right) \tag{4.9}$$

$$CaCO_3 + 2HCOOH \rightarrow Ca\left(HCOO\right)_2 + H_2O + CO_2\uparrow\left(\Delta G = -5.59\,kJ/mol\right) \tag{4.10}$$

$$MgCO_3 + 2HCOOH \rightarrow Mg\left(HCOO\right)_2 + H_2O + CO_2\uparrow\left(\Delta G = -4.83\,kJ/mol\right) \tag{4.11}$$

These reactions demonstrate the typical behavior of carbonate minerals in acidic media, where carbonate ions (CO_3^{2-}) undergo neutralization by protonic hydrogen ions (H^+) from the acid, leading to the evolution of carbon dioxide gas. The visible manifestation of this gas evolution—effervescence or bubbling—indicates the progression of the reaction.

In the case of citric acid, the reaction leads to the formation of calcium citrate [$Ca_3(C_6H_5O_7)_2$] and magnesium citrate [$Mg_3(C_6H_5O_7)_2$], while formic acid yields calcium formate [$Ca(HCOO)_2$] and magnesium formate [$Mg(HCOO)_2$]. These products are generally soluble in aqueous solution, enabling efficient separation from the solid matrix.

Mechanistically, these reactions represent a double displacement process (also described as metathesis reactions), in which the metal cations (Ca^{2+} and Mg^{2+}) are substituted for hydrogen ions from the acid. This leads to the formation of new, soluble organometallic salts while releasing carbon dioxide—a feature that not only drives the reaction forward thermodynamically (as reflected in the negative ΔG values) but also aids in the physical detection of reaction progress through gas release.

The negative values of Gibbs free energy (ΔG) associated with these reactions, particularly in the case of citric acid, confirm the spontaneous and thermodynamically favorable nature of the leaching process under standard conditions. The significantly more negative ΔG for the magnesium citrate formation suggests a greater reactivity of magnesium carbonate in comparison to calcium carbonate when treated with citric acid.

Both citric acid and formic acid are classified as weak organic acids by nature, but they differ significantly in their molecular structures and acid-base behaviors, which in turn affect their reactivity with carbonate minerals such as calcium carbonate ($CaCO_3$) and magnesium carbonate ($MgCO_3$).

Citric acid ($C_6H_8O_7$) is a triprotic carboxylic acid, meaning it possesses three dissociable carboxyl groups (–COOH) in its molecular structure. In addition, it contains a hydroxyl group (–OH), which increases the molecule's polarity and enhances its aqueous solubility. The presence of three acidic protons enables citric acid to act not only as a buffering agent but also as an effective complexing ligand in metal-ion coordination processes. This polyprotic nature imparts greater chemical versatility, especially in reactions involving metal ions, as it can form stable chelate complexes [144].

In contrast, formic acid (HCOOH) is the simplest monocarboxylic acid, containing only one carboxyl group and no additional functional moieties. Structurally, it is composed of only two carbon atoms, which makes it less polar compared to citric acid. Despite its simplicity, formic acid exhibits strong acidity for a monobasic acid, although its ability to form metal complexes is limited due to the absence of multiple donor sites.

The difference in the acids' dissociation capacity and functional groups directly influences their reactivity with metal carbonates. Citric acid, being triprotic, can release three protons, contributing to its ability to dissolve carbonates more effectively and to form stable metal-citrate complexes. However, the presence of multiple carboxylate groups also leads to charge delocalization, which moderates the strength of individual acidic protons compared to monocarboxylic acids like formic acid.

The thermodynamic parameters associated with the carbonate-acid reactions highlight their differential energy profiles. For instance:

- In reaction (4.8), calcium carbonate reacts with citric acid yielding a Gibbs free energy change (ΔG) of −56 kJ/mol, indicating a mildly exergonic process. This suggests the reaction proceeds spontaneously under standard conditions with moderate energy release, forming calcium citrate and releasing CO_2.

- In reaction (4.9), the interaction between magnesium carbonate and citric acid results in a significantly more negative ΔG of −117 kJ/mol, reflecting a higher thermodynamic favorability compared to calcium. This indicates that magnesium carbonate dissolves more readily in citric acid under the same conditions.
- Conversely, the reactions with formic acid display substantially lower thermodynamic efficiency. In reaction (4.10), $CaCO_3$ yields a ΔG of only −5.59 kJ/mol, indicating that the process is barely exergonic, and may proceed only partially or slowly without additional driving forces (e.g., elevated temperature or agitation).
- Similarly, reaction (4.11) involving $MgCO_3$ and formic acid results in a ΔG of −4.83 kJ/mol, again suggesting limited spontaneity and weaker driving force compared to the citric acid reactions.

These results indicate that citric acid is significantly more effective than formic acid in the selective leaching of calcium and magnesium carbonates. The greater exergonicity of reactions with citric acid reflects its stronger chelating ability, higher number of dissociable protons, and capacity to form stable, water-soluble metal-organic complexes (citrates).

On the other hand, the weak energy yield in formic acid reactions suggests that additional process conditions—such as elevated temperature, pressure, or acid concentration—may be required to achieve comparable dissolution efficiency.

In summary, from a thermodynamic perspective, citric acid offers a superior leaching agent for carbonate decomposition in phosphate raw materials, particularly when targeting the removal of calcium and magnesium phases. These findings support the selection of citric acid in eco-friendly, low-temperature leaching systems for mineral beneficiation.

4.2 Use of Acetic Acid for the Beneficiation of Low-Grade Phosphorites

The experimental investigations conducted on the beneficiation of phosphate-siliceous layered rocks using acetic acid are summarized and presented in Tables 4.8 and 4.9 and Fig. 4.9. These results provide insight into the efficiency of acetic acid leaching for selective dissolution of non-phosphate minerals and the concentration of valuable phosphate components under controlled laboratory conditions.

The data presented in Fig. 4.9 reflect the dynamic behavior of phosphate-siliceous rock leaching when treated with acetic acid under controlled variations of temperature and leaching duration. The parameters measured include the pH of the system, the concentration of leached phosphorus (C, mol/L), the leaching rate (ν, mol/L·min), and the degree of phosphorus extraction (α, %).

A gradual increase in pH values was observed over time at all temperature levels (293, 303, and 313 K), indicating a weakening of the acidic medium as the

Table 4.8 Experimental data on the beneficiation of phosphate-siliceous layered rocks with acetic acid

Temperature, K	Time, min	pH	*C*, mol/L	ν, mol/L·min	α, %
293	10	3.251	0.00056104	0.20196	88.00
	20	3.263	0.00054575	0.10136	88.33
	30	3.275	0.00053088	0.06781	88.64
	40	3.282	0.00052239	0.05096	88.83
	50	3.297	0.00050466	0.04094	89.21
303	10	3.329	0.00046881	0.20648	89.97
	20	3.332	0.00046558	0.10332	90.04
	30	3.358	0.00043853	0.06932	90.62
	40	3.377	0.00041975	0.05222	91.02
	50	3.380	0.00041686	0.04180	91.08
313	10	3.391	0.00040644	0.20956	91.31
	20	3.397	0.00040086	0.10490	91.42
	30	3.401	0.00039719	0.07000	91.50
	40	3.406	0.00039264	0.05255	91.60
	50	3.414	0.00038547	0.04211	91.75

Table 4.9 Technological parameters of the beneficiation process of phosphate-siliceous layered rocks using acetic acid

Temperature, K	Time, min	P_2O_5, %	Percentage of P_2O_5 dissolved in solution	Enrichment level of P_2O_5, %	*E*, %	*K*
293	10	13.72	–	54.19	53.11	1.14
	20	14.70	–	55.58	53.86	1.22
	30	15.07	–	56.22	54.24	1.25
	40	15.03	–	56.07	54.12	1.25
	50	15.08	–	55.79	53.82	1.26
303	10	20.07	–	74.66	67.79	1.68
	20	20.01	0.23	75.49	68.60	1.67
	30	19.94	0.92	74.40	67.67	1.66
	40	19.57	1.08	73.01	66.71	1.63
	50	19.32	1.26	71.15	65.21	1.61
313	10	19.55	1.21	68.99	63.05	1.63
	20	19.75	1.24	69.48	63.34	1.64
	30	19.31	1.33	68.84	63.10	1.61
	40	19.21	1.57	68.66	63.02	1.60
	50	19.18	1.61	66.97	61.49	1.60

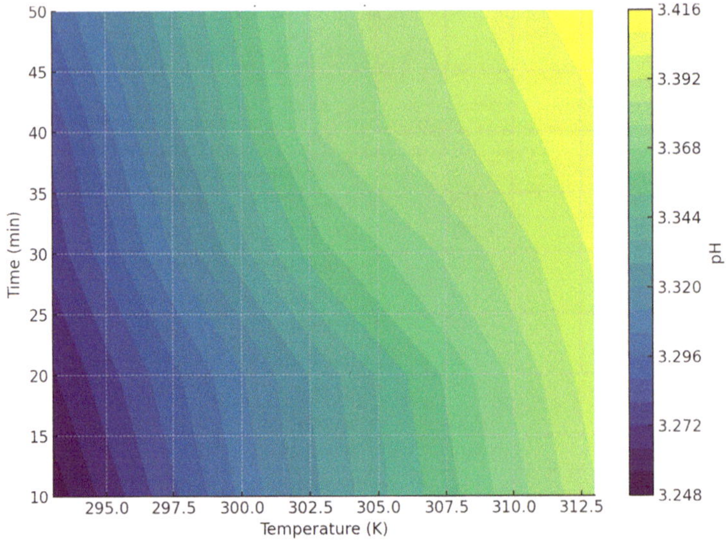

Contour Plot of pH vs Temperature and Time

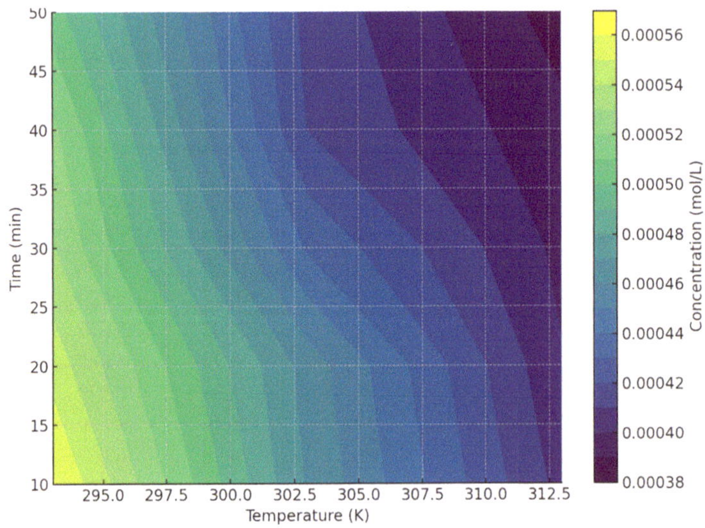

Contour Plot of Concentration vs. Temperature and Time

Fig. 4.9 Experimental data on the beneficiation of phosphate-siliceous layered rocks with acetic acid

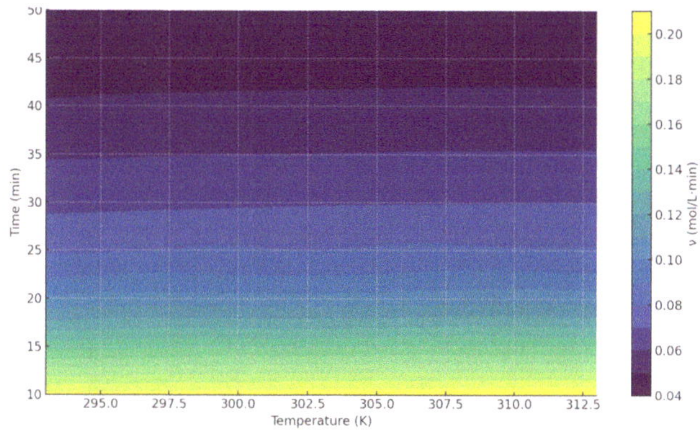

Contour Plot of Reaction Rate v as Function of Temperature and Time

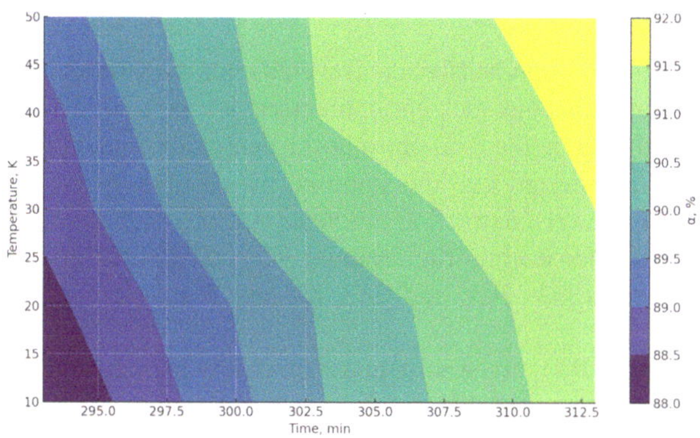

Contour plot showing the dependency of the conversion rate (α, %) on both temperature (K) and time (minutes)

Fig. 4.9 (continued)

dissolution progresses. At 293 K, the pH increases from 3.251 at 10 min to 3.297 at 50 min, while at 313 K, it increases from 3.391 to 3.414 over the same interval. This rise in pH corresponds to the neutralization of acetic acid due to ion exchange and leaching reactions involving the carbonate and phosphate minerals.

The concentration of phosphorus in solution decreases over time, regardless of temperature, indicating that the leaching process is most intensive in the initial stage. For instance, at 293 K, the concentration drops from 0.00056104 mol/L at 10 min to 0.00050466 mol/L at 50 min. This trend is consistent with saturation effects and reduced reactivity as the easily soluble phases are depleted.

Temperature exerts a clear influence on dissolution: at the initial 10-min mark, the phosphorus concentration is higher at lower temperatures, but this relationship diminishes with time due to differences in reaction kinetics and diffusion control.

The rate of leaching (ν) shows an exponential decay with time, emphasizing the kinetic control of the process. The highest leaching rate is observed in the first 10 min across all temperatures (e.g., 0.20196 mol/L·min at 293 K and 0.20956 mol/L·min at 313 K), after which the rate steadily declines. This suggests that the initial acid attack on reactive mineral surfaces is rapid, followed by a slower diffusion-controlled regime [145, 146].

Notably, despite a decreasing trend in concentration (C), the leaching rate remains slightly higher at elevated temperatures, confirming the role of thermal activation in promoting dissolution kinetics.

The phosphorus extraction efficiency (α) increases both with temperature and time. At 293 K, α improves modestly from 88.00% to 89.21% over 50 min. At 313 K, the extraction reaches 91.75%, indicating that even small temperature increases yield significantly higher efficiency, likely due to enhanced molecular mobility and acid reactivity. The convergence of high α values at longer durations across all temperatures implies that while time enhances the completeness of the process, temperature accelerates it.

This Fig. 4.10 presents a comprehensive dataset characterizing the technological performance of phosphate-siliceous rock enrichment via leaching with acetic acid. The indicators evaluated include the total phosphorus pentoxide content (P_2O_5, %), the percentage of P_2O_5 transferred into the solution, the relative increase in P_2O_5 concentration (enrichment level), extraction efficiency (E, %), and the enrichment ratio (K). These metrics offer a multi-dimensional view of process effectiveness across three temperatures (293, 303, and 313 K) and five leaching durations (10–50 min).

The measured content of P_2O_5 in the solid phase gradually increases with time at all temperatures, indicating successful leaching and gradual enrichment. At 293 K, the P_2O_5 content rises from 13.72% to 15.08%, while at 303 K it reaches a maximum of 20.07% after only 10 min. This sharp contrast suggests that temperature significantly enhances the reactivity and solubility of phosphorus-bearing minerals in acetic acid.

However, at 313 K, although the P_2O_5 content starts at a relatively high value (19.55%), it shows only marginal improvement over time, and even a slight decline beyond 20 min, indicating possible re-precipitation or saturation effects.

The parameter "P_2O_5 transferred into the solution" reflects the portion of total phosphorus solubilized. Notably, this parameter is not reported at 293 K, possibly due to negligible or undetectable dissolution. Starting from 303 K, quantifiable amounts appear—beginning at 0.23% at 20 min and rising to 1.26% at 50 min. At 313 K, the values range from 1.21% to 1.61%, showing slightly greater transfer, though the increase with time is less pronounced.

This suggests that while higher temperatures favor more efficient dissolution, the majority of phosphorus is retained in the solid phase—likely due to partial recrystallization or formation of sparingly soluble complexes.

The enrichment level (ΔP_2O_5) illustrates the percentage gain in phosphorus concentration relative to the starting material. This indicator increases steadily with time at all temperatures, peaking at 74.66% at 303 K (10 min) and reaching 68.99%

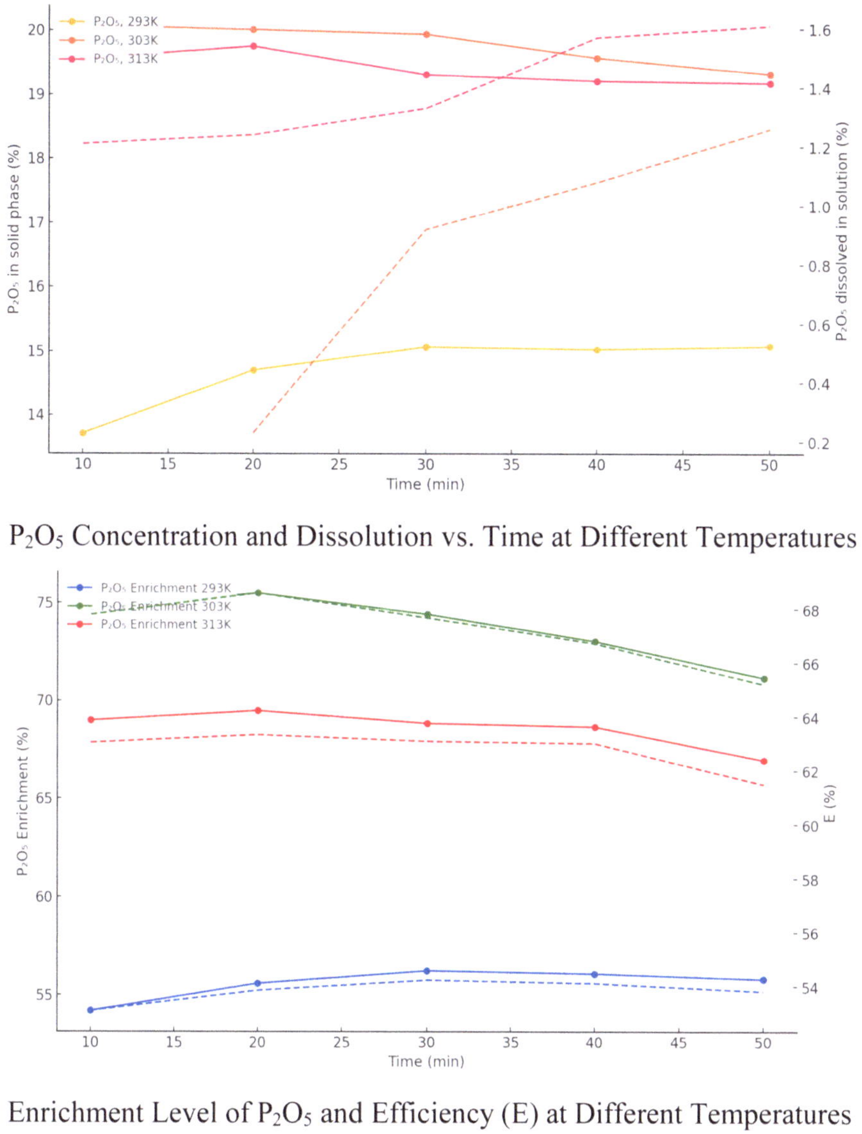

Fig. 4.10 Technological indicators of the enrichment process of phosphate-siliceous rocks using acetic acid

at 313 K (10 min). With further time, a slight decrease in enrichment is noted, pointing to the diffusion limitation or completion of easily accessible phosphate phase dissolution [146].

Extraction efficiency is a critical metric reflecting the percentage of phosphorus recovered relative to its original quantity. The highest value of 68.60% is recorded at 303 K after 20 min. However, a gradual decline is observed with time at both 303 and 313 K, suggesting that extended leaching may not enhance extraction due to equilibrium limitations or side reactions. At 293 K, the maximum efficiency reached is 54.24%, confirming the limited effectiveness of leaching at lower temperatures.

The enrichment ratio (*K*) represents the ratio of final to initial P_2O_5 content in the solid phase. At 293 K, values increase from 1.14 to 1.26, indicating modest but consistent enrichment. At 303 K, the ratio rises sharply to 1.68, affirming the superior efficiency of leaching at this intermediate temperature. Interestingly, the values at 313 K remain relatively stable around 1.60–1.63, supporting the notion that 303 K may represent the optimal thermal regime for balancing reactivity and selectivity.

The data from Fig. 4.10 support the conclusion that temperature is the dominant factor affecting the enrichment efficiency of phosphate-siliceous rocks with acetic acid. A temperature of 303 K offers the best combination of high phosphorus content, significant solubilization, high enrichment level, and extraction efficiency. Although further heating to 313 K slightly increases solubilized phosphorus, the overall benefit plateaus due to diminishing returns in extraction and a possible decline in product quality. These findings are essential for optimizing eco-friendly and low-energy phosphate beneficiation protocols.

The results of experimental studies on the enrichment of low-grade phosphorites from the Akzhar deposit using acetic acid are presented in Tables 4.10 and 4.11, as well as in Fig. 4.11.

Table 4.10 presents a comprehensive set of experimental results concerning the leaching of low-grade phosphorites from the Akzhar deposit with acetic acid under varying temperatures (293, 303, and 313 K) and contact times (10–50 min). The studied parameters include the pH of the leach solution, concentration of soluble phosphorus in the form of P-containing species (*C*, mol/L), dissolution rate (ν, mol/L·min), and degree of phosphorus recovery (α, %).

Across all temperature conditions, a gradual increase in pH values is observed with increasing leaching time, reflecting the progressive neutralization of acetic acid due to interaction with phosphate minerals. At 293 K, the initial pH at 10 min was 3.452, increasing to 3.533 at 50 min. A similar trend is observed at 303 and 313 K, with final pH values of 3.545 and 3.611, respectively, suggesting enhanced buffering capacity and reduced acid consumption at higher temperatures.

The concentration of phosphorus in solution (*C*) consistently decreased with time, indicative of the depletion of easily soluble phosphorus species and the deceleration of the leaching process. For instance, at 293 K, *C* decreased from 0.00035318 mol/L at 10 min to 0.00029308 mol/L at 50 min. This corresponds to a decline in dissolution rate (ν), which dropped from 0.18300 to 0.03711 mol/L·min

Table 4.10 Experimental data on the enrichment of low-grade phosphorites from the Akzhar deposit using acetic acid

Temperature, *K*	Time, min	pH	*C*, mol/L	ν, mol/L·min	α, %
293	10	3.452	0.00035318	0.18300	92.44
	20	3.487	0.00032583	0.09208	93.03
	30	3.515	0.00030549	0.06167	93.46
	40	3.528	0.00029648	0.04635	93.66
	50	3.533	0.00029308	0.03711	93.73
303	10	3.538	0.00028973	0.18569	93.80
	20	3.540	0.00028840	0.09287	93.83
	30	3.542	0.00028707	0.06193	93.86
	40	3.544	0.00028575	0.04646	93.89
	50	3.545	0.00028510	0.03717	93.90
313	10	3.567	0.00027101	0.18648	94.20
	20	3.570	0.00026915	0.09328	94.24
	30	3.584	0.00026061	0.06230	94.42
	40	3.597	0.00025292	0.04681	94.59
	50	3.611	0.00024490	0.03751	94.76

Table 4.11 Technological indicators of the enrichment process of low-grade phosphorites from the Akzhar deposit using acetic acid

Temperature, K	Time, min	P_2O_5, %	Percentage of P_2O_5 dissolved in solution	Enrichment level of P_2O_5, %	*E*, %	*K*
293	10	18.54	–	56.71	54.16	1.26
	20	18.86	–	57.77	54.96	1.28
	30	19.28	–	58.64	55.50	1.31
	40	19.13	–	58.45	55.42	1.29
	50	19.04	–	58.32	55.36	1.29
303	10	20.16	–	63.54	59.48	1.37
	20	20.73	–	64.33	59.79	1.40
	30	20.86	–	65.04	60.35	1.41
	40	20.82	–	65.20	60.53	1.41
	50	20.72	0.19	65.11	60.52	1.40
313	10	21.43	0.37	68.71	63.30	1.45
	20	21.85	0.41	69.06	63.28	1.48
	30	22.11	0.34	70.82	64.68	1.50
	40	22.21	0.37	71.56	65.27	1.51
	50	22.18	0.32	70.67	64.48	1.50

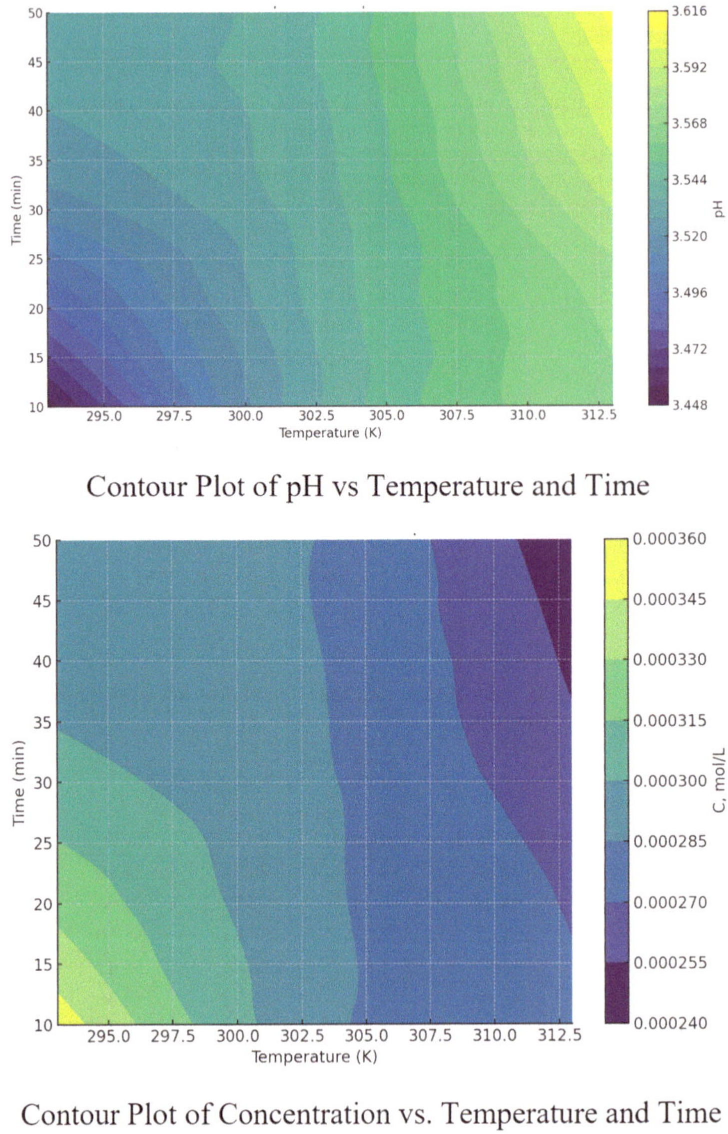

Fig. 4.11 Experimental data on the enrichment of low-grade phosphorites from the Akzhar deposit using acetic acid

over the same time frame. Such reductions in ν are attributed to the approach toward leaching equilibrium and surface passivation phenomena.

The degree of phosphorus recovery (α) demonstrates a steady increase with both temperature and time. At 293 K, α rose from 92.44% to 93.73%; at 303 K, from 93.80% to 93.90%; and at 313 K, from 94.20% to 94.76%. These results suggest that higher temperatures enhance both the kinetics and thermodynamics of the

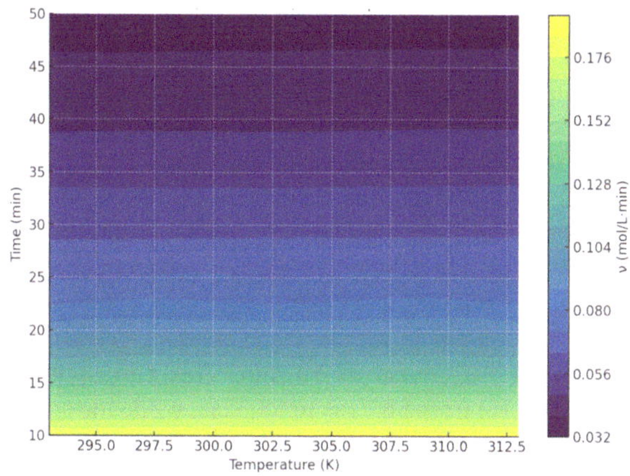

Contour Plot of Reaction Rate ν as Function of Temperature and Time

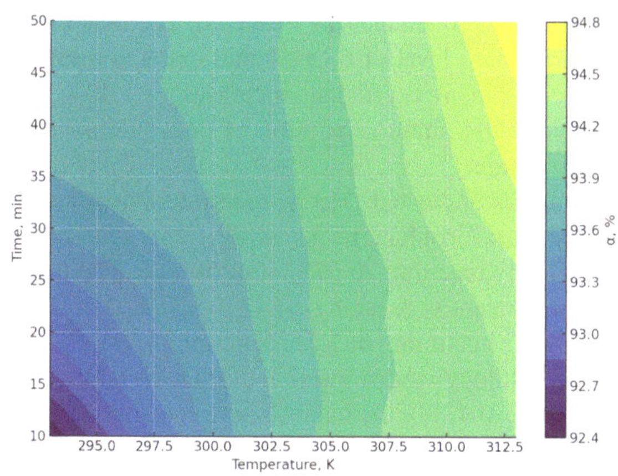

Contour plot showing the dependency of the conversion rate (α, %) on both temperature (K) and time (minutes)

Fig. 4.11 (continued)

leaching reaction, enabling more efficient extraction of phosphorus even at relatively low acid concentrations.

Importantly, the narrow variation of pH and concentration values at 303 and 313 K over time indicates that the system rapidly reaches a quasi-steady state. This stability reflects the robustness of the acetic acid leaching mechanism under optimized conditions, offering favorable selectivity and high extraction efficiency for low-grade phosphorites.

In summary, the data from Table 4.10 confirm that acetic acid is an effective leaching agent for the beneficiation of the Akzhar phosphorite ore. Optimal enrichment parameters are achieved at 313 K and 50 min of leaching time, resulting in a high phosphorus recovery rate ($\alpha = 94.76\%$) with minimal reagent consumption and stable process conditions.

Table 4.11 presents the technological performance metrics of the enrichment process for low-grade phosphorites from the Akzhar deposit using acetic acid as the leaching agent. The dataset includes variations in temperature (293–313 K), leaching duration (10–50 min), phosphorus content in the solid phase after treatment (P_2O_5, %), the proportion of P_2O_5 dissolved into the solution, the enrichment level of P_2O_5, phosphorus extraction efficiency (E, %), and enrichment coefficient (K) (Fig. 4.12).

At 293 K, the P_2O_5 content in the enriched product gradually increases from 18.54% to 19.04% over the course of 50 min. This corresponds to a moderate increase in enrichment level from 56.71% to 58.32%, while the extraction efficiency (E) rises from 54.16% to 55.36%. The enrichment coefficient (K), which reflects the ratio of phosphorus concentration in the concentrate to that in the original ore, ranges from 1.26 to 1.29, indicating steady but limited phosphorus upgrading at this baseline temperature.

At 303 K, significant improvements are observed. The final P_2O_5 content reaches 20.72%, with an enrichment level of 65.11% and extraction efficiency of 60.52% after 50 min. The enrichment coefficient increases to 1.40–1.41, which reflects enhanced solubilization and phase separation of phosphorus-bearing components due to elevated temperature.

At the highest temperature tested, 313 K, the system demonstrates optimal performance. The phosphorus content peaks at 22.18%, and the enrichment level exceeds 70%, specifically reaching 70.67% after 50 min. The extraction efficiency (E) also improves significantly, reaching 64.48–65.27%. Notably, the enrichment coefficient (K) achieves a maximum of 1.51, denoting superior phosphorus selectivity and concentration. Although the proportion of P_2O_5 dissolved in solution remains relatively low (e.g., 0.32–0.41%), the balance between solid-phase enrichment and solution-phase losses remains favorable.

These results confirm the thermally enhanced nature of acetic acid leaching. The increase in temperature facilitates deeper penetration of the leachate, improved dissolution kinetics, and stabilization of beneficial reaction intermediates. Importantly, the increase in P_2O_5 content with minimal solution loss suggests that acetic acid preferentially reacts with non-phosphate gangue minerals, thereby enriching the solid phase while minimizing phosphorus depletion into the liquid phase.

The results of the experimental studies on the acetic-acid beneficiation of low-grade phosphorites from the Shilisai deposit are summarized in Tables 4.12 and 4.13 and illustrated in Fig. 4.13.

Table 4.12 (reproduced above) records the time-resolved evolution of pH, residual free-acid concentration C, instantaneous leaching rate ν, and the fractional acid utilization α during the acetic-acid treatment of the Shilisai ore at 293–313 K. A systematic increase of T and τ promotes a coordinated shift in all four process variables, revealing clear kinetic regimes that are instructive for industrial design.

At every temperature the highest ν ($\approx$0.21 mol L^{-1} min^{-1}) is registered in the first 10 min, when fresh acid contacts readily accessible calcite/dolomite surfaces. The

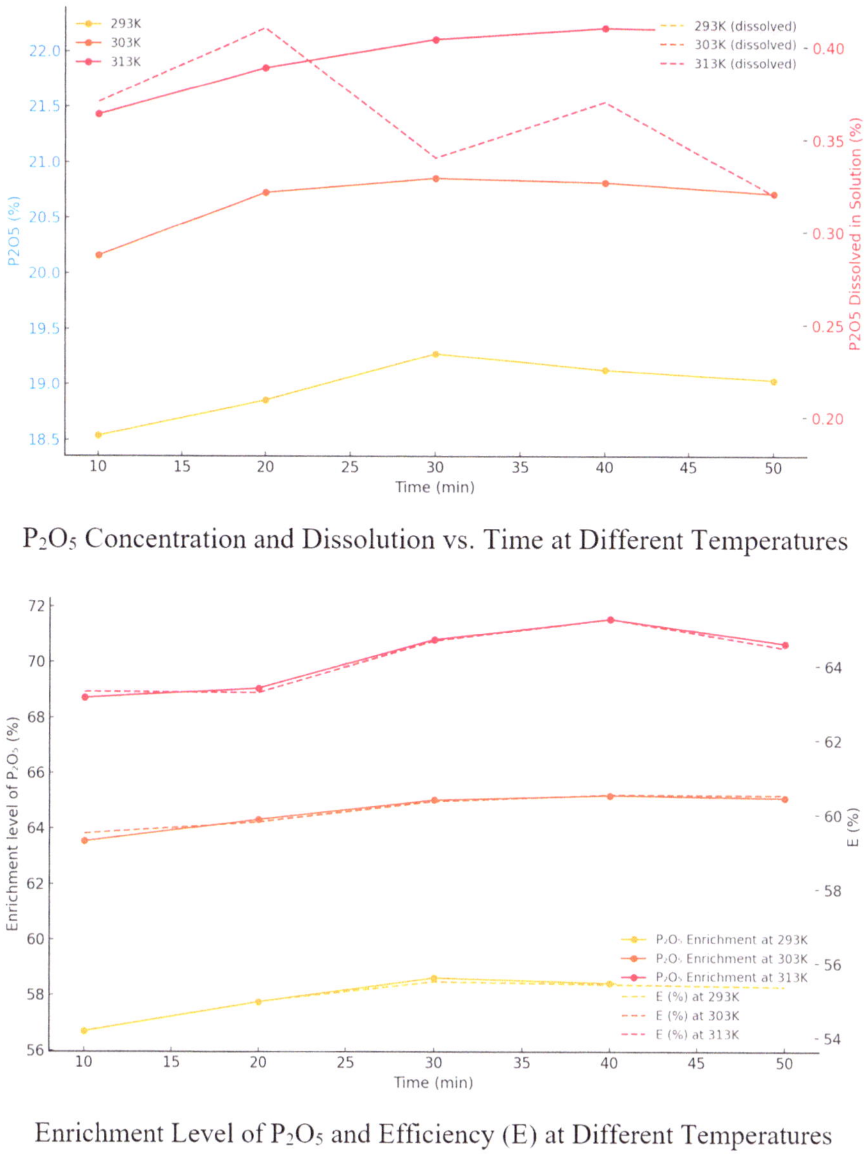

Fig. 4.12 Technological indicators of the enrichment process of low-grade phosphorites from the Akzhar deposit using acetic acid

Table 4.12 Experimental data on the enrichment of low-grade phosphorites from the Shilisai deposit using acetic acid

Temperature, K	Time, min	pH	*C*, mol/L	ν, mol/L·min	α, %
293	10	2.877	0.0013273	0.21354	87.61
	20	2.880	0.0013182	0.10687	87.69
	30	2.884	0.0013061	0.07134	87.81
	40	2.891	0.0012852	0.05362	88.00
	50	2.897	0.0012676	0.04297	88.16
303	10	2.898	0.0012647	0.21496	88.19
	20	2.900	0.0012589	0.10755	88.25
	30	2.905	0.0012445	0.07180	88.38
	40	2.912	0.0012246	0.05397	88.57
	50	2.925	0.0011885	0.04333	88.90
313	10	2.931	0.0011721	0.21708	89.06
	20	2.943	0.0011402	0.10889	89.35
	30	2.952	0.0011168	0.07277	89.57
	40	2.983	0.0010399	0.05502	90.29
	50	3.016	0.0009638	0.04436	91.00

Table 4.13 Technological indicators of the enrichment process of low-grade phosphorites from the Shilisai deposit using acetic acid

Temperature, K	Time, min	P_2O_5, %	Percentage of P_2O_5 dissolved in solution	Enrichment level of P_2O_5, %	*E*, %	*K*
293	10	19.73	–	49.41	48.07	1.12
	20	19.89	–	52.15	50.63	1.13
	30	20.08	–	53.22	51.55	1.14
	40	20.13	–	54.27	52.53	1.15
	50	20.04	–	54.11	52.44	1.14
303	10	23.16	–	62.84	58.52	1.32
	20	24.73	–	63.99	58.38	1.41
	30	25.41	–	65.58	59.29	1.45
	40	25.76	–	66.60	59.93	1.47
	50	24.72	1.12	65.21	59.50	1.41
313	10	22.03	1.53	55.71	52.65	1.25
	20	22.85	1.76	56.14	52.49	1.30
	30	23.11	2.19	58.10	54.14	1.32
	40	23.01	2.58	57.88	54.01	1.31
	50	22.78	2.91	56.03	52.44	1.30

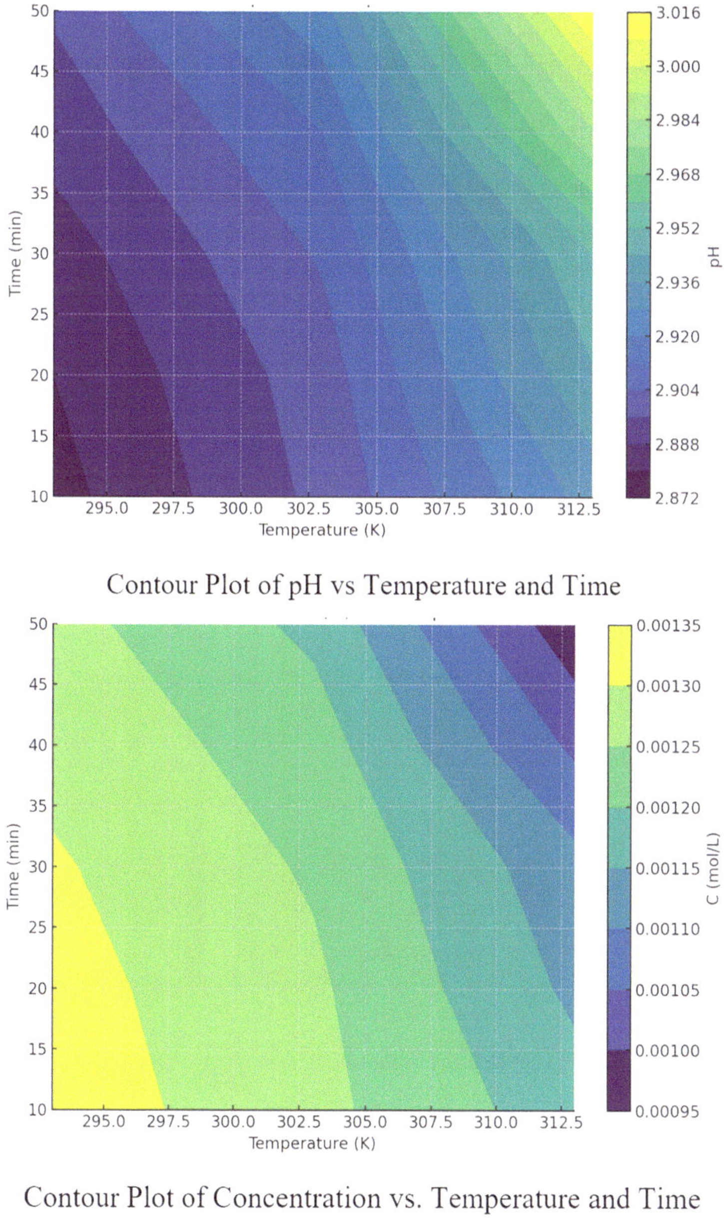

Fig. 4.13 Experimental data on the enrichment of low-grade phosphorites from the Shilisai deposit using acetic acid

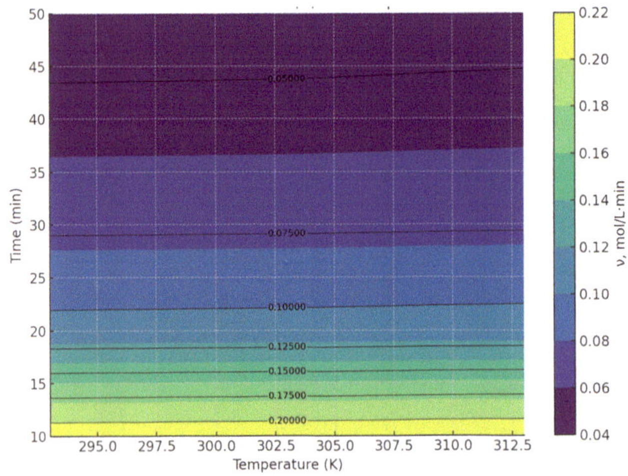

Contour Plot of Reaction Rate ν as Function of Temperature and Time

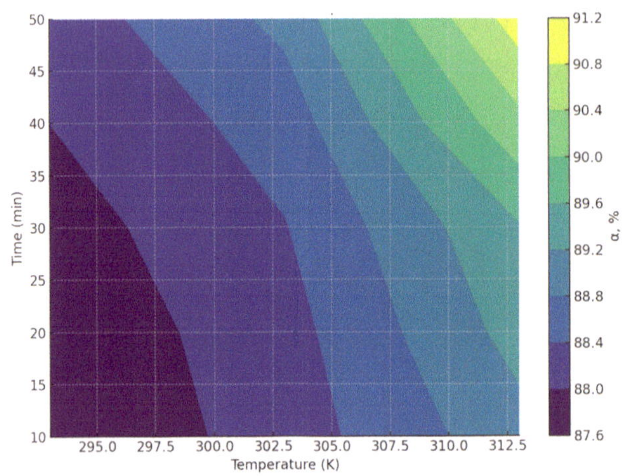

Contour plot showing the dependency of the conversion rate (α, %) on both temperature (K) and time (minutes)

Fig. 4.13 (continued)

concomitant decline in C (2.5–3.5%) and a pH rise of ~0.02 units confirm that proton attack is initially surface-reaction controlled. Because ν shows only a marginal sensitivity to temperature during this period (0.213 at 293 K vs. 0.217 mol L^{-1} min^{-1} at 313 K).

Beyond 20 min, ν drops by ~50% while pH continues to climb and C falls further, signaling that the reaction front progresses into partially reacted grains. The reduced dependence of ν on temperature in this interval implies that intragranular diffusion of acetic acid (and/or outward CO_2 venting) now limits the overall rate.

After 50 min, the utilization coefficient α reaches 88–91%. Raising the temperature from 293 to 313 K lifts α by 2.9 percentage points, yet consumes only ~5% additional acid. Economically, operating at 303 K for 40 min (pH $\approx$2.91, $C \approx 1.22 \times 10^{-3}$ mol L^{-1}, $\alpha \approx$88.6%) represents a pragmatic compromise between reagent efficiency and energy input; moving to 313 K is justified only when maximum phosphorus recovery is mandatory.

A 30–40 min hold satisfies >95% of the attainable α at any temperature, indicating that longer runs add minimal benefit. A modest 10 K elevation (293 → 303 K) yields a disproportionate 0.7–0.9% rise in α and 11% higher P_2O_5 upgrade (see Table 4.12), underscoring temperature as a key lever for throughput rather than kinetics. Because α never exceeds 91%, full conversion of carbonate is unattainable under the chosen L:S = 3:1 ratio; optimization could explore slightly higher acid charges or staged addition to push α closer to 95% without large capital penalties.

Overall, the data validate the suitability of acetic acid for selective decarbonation of Shilisai phosphorite: even at the lowest temperature examined, more than 87% of the dosed acid is effectively consumed, and the decarbonated residue concentrates P_2O_5 by a factor $K \approx 1.5$. These kinetic insights provide the quantitative basis for scaling the process to a continuous, stirred reactor configuration operating at ~303 K and 30 min mean residence time.

The kinetic-technological profile obtained for low-grade phosphorites from the Shilisai deposit treated with acetic acid reveals a clear temperature-dependent hierarchy of performance. At 293 K the leach liquor remains virtually free of dissolved phosphate throughout the entire 40- to 50-min interval, confirming that the acid reacts almost exclusively with the carbonate gangue. As the carbonates are progressively removed the P_2O_5 concentration in the solid rises from an initial 19.7% to just above 20%, and the enrichment level advances to roughly 54%, while the overall efficiency stabilizes at about 52%. Because only a modest fraction of the acid is consumed at this temperature, the concentration ratio K remains close to unity and the degree of underutilized reagent indicates an incomplete decarbonation (Fig. 4.14).

When the temperature is increased to 303 K the system passes into a markedly more efficient regime. Within the first half hour the P_2O_5 grade climbs to 25–26%, enrichment exceeds 66% and the concentration ratio reaches 1.47. Throughout the first 40 min the leachate still contains no analytically detectable phosphate, demonstrating that the selective window of the acid remains open under these conditions. Only after 50 min does a small but measurable fraction of P_2O_5 appear in solution, signaling the onset of apatite attack and thus defining the upper boundary of the

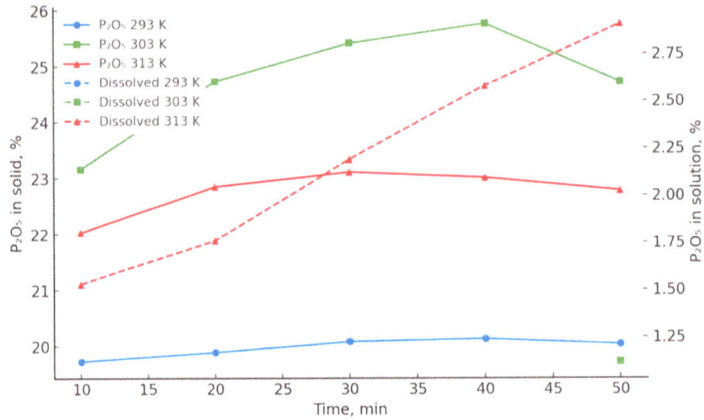

P_2O_5 Concentration and Dissolution vs. Time at Different Temperatures

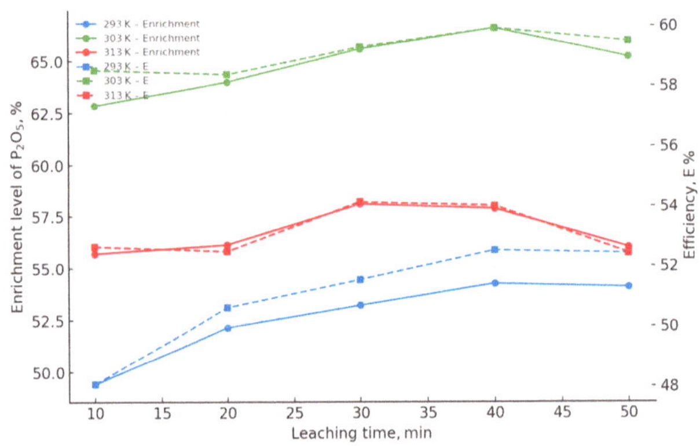

Enrichment Level of P_2O_5 and Efficiency (E) at Different Temperatures

Fig. 4.14 Technological indicators of the enrichment process of low-grade phosphorites from the Shilisai deposit using acetic acid

optimal residence time at this temperature. At the same point the concentrate grade shows a slight decline, corroborating that continued leaching beyond the window of selectivity yields no further beneficiation benefit.

Operation at 313 K shifts the reaction balance in favor of nonselective dissolution. Immediately after 10 min aliquots contain more than one and a half per cent of the total phosphorus in soluble form, and this proportion continues to rise with time. Although carbonate removal is faster—reflected in an acid-consumption degree (α) greater than 90%—the concomitant loss of phosphate limits the concentrate grade to approximately 23% P_2O_5. The enrichment factor retreats accordingly, and the process efficiency falls below the values recorded at 303 K despite the higher intrinsic reaction rate. Thermodynamically the system approaches a regime in which further temperature increments merely accelerate the decomposition of apatite without appreciably enhancing carbonate elimination.

Table 4.14 Optimized process parameters identified for acetic acid beneficiation

Feedstock	S/L ratio	P_2O_5 in concentrate, %	Temperature, °C	Residence time, min	P_2O_5 ED, %	Process efficiency, *E* %	Concentration ratio, *K*
Phosphate–siliceous bed-rock (tailings)	1:3	20.04	20–30	20–30	75.07	68.19	1.67
Akzhar low-grade phosphorite	1:3	22.19	30–40	30–40	71.19	64.77	1.50
Shilisai low-grade phosphorite	1:3	25.58	30–40	30–40	66.09	59.71	1.46

The collective evidence therefore points to 303 K and a reaction time of 35–40 min as the most favorable operating window for single-stage acetic-acid beneficiation of Shilisai ore. Under these conditions the acid is utilized to about 89%, carbonates are almost completely removed, and phosphate losses remain negligible. Temperatures above this threshold accelerate apatite dissolution and erode concentrate quality, whereas lower temperatures lengthen the residence time required to reach an equivalent enrichment level. Process scale-up should consequently emphasize tight thermal control around the identified optimum and employ on-line pH monitoring to terminate the leach as soon as the liquor approaches pH ≈2.90, thereby preserving selectivity and minimizing reagent wastage.

The consolidated optimum operating parameters derived for all three feedstocks are summarized in Table 4.14.

The parameter matrix in Table 4.14 establishes a common liquid–solid ratio of 1:3 as adequate for quantitative dissolution of the calcite–dolomite gangue while avoiding excessive dilution of the acetate liquor. Under these conditions: phosphate–siliceous tailings (essentially a technogenic waste stream rich in quartz) attain an upgrading degree of 75% within only 20–30 min at 20–30 °C, confirming that low-temperature leaching is sufficient once fine grinding exposes the carbonate intergrowths. Akzhar ore, characterized by a lower carbonate buffer capacity but higher clay content, requires a modest increase in temperature (30–40 °C) and time (30–40 min) to stabilize the apatite phase against re-adsorption of dissolved Ca^{2+}/Mg^{2+}. The enrichment degree of 71% and an overall efficiency of 65% compare favorably with literature data for ores of similar mineralogy. Shilisai ore contains the highest initial P_2O_5 yet also the greatest dolomite fraction. Operating in the same 30–40 °C window yields a concentrate grading 25.6% P_2O_5—adequate for direct wet-process acidulation—and a concentration ratio of 1.46.

Earlier investigations on sub-economic phosphorites and carbonate-rich tailings report optimal acetic-acid leaching at 40–60 min and 20–70 °C [105, 107, 114, 121, 147]. Extending the present experiments beyond 40 min or above 40 °C produced no additional gain in P_2O_5 grade or mass yield, but did increase reagent consumption and calcium acetate loading in the recycle loop. Hence the shorter, milder conditions recommended here not only reduce energy demand but also ease subsequent acid regeneration.

The close agreement between the present dataset and published kinetics for Kazakhstani deposits corroborates the reproducibility of acetic-acid beneficiation across geological provinces and underscores its suitability for valorizing both run-of-mine low-grade ores and legacy tailings.

4.2.1 Kinetic Treatment of the Experimental Data

Heterogeneous liquid–solid reactions play a key role in many chemical and hydrometallurgical operations, and the successful design and scale-up of the corresponding reactors relies on accurate kinetic information. To elucidate the rate-controlling step and derive intrinsic kinetic parameters for the selective dissolution of carbonate phases from phosphorites in acetic acid, the experimental leaching data were analyzed with the heterogeneous-reaction model introduced in Eq. (3.14).

The goodness of fit and the internal consistency of the derived parameters were verified by complementary statistical tests and by conventional graphical methods. Results of the kinetic treatment for the three representative feedstocks—phosphate–siliceous bed rock, and the low-grade phosphorites from the Akzhar and Shilisai deposits—are compiled in Tables 4.15, 4.16 and 4.17, respectively, and are summarized graphically in Fig. 4.15.

Table 4.15 Results of the kinetic analysis of experimental data obtained from the beneficiation process of phosphate–siliceous layered rocks (Zhanatas deposit)

Temperature, K	τ, min	$\sqrt{\tau}$	α (single shares)	$1-\alpha$	$\sqrt[3]{1-\alpha}$	$1-\sqrt[3]{1-\alpha}$
293	10	3.16	0.8800	0.1200	0.493242	0.506758
	20	4.47	0.8833	0.1167	0.488679	0.511321
	30	5.47	0.8864	0.1136	0.484313	0.515687
	40	6.32	0.8883	0.1117	0.481598	0.518402
	50	7.07	0.8921	0.1079	0.476073	0.523927
303	10	3.16	0.8997	0.1003	0.464623	0.535377
	20	4.47	0.9004	0.0996	0.463539	0.536461
	30	5.47	0.9062	0.0938	0.454361	0.545639
	40	6.32	0.9102	0.0898	0.447808	0.552192
	50	7.07	0.9108	0.0892	0.446809	0.553191
313	10	3.16	0.9131	0.0869	0.442935	0.557065
	20	4.47	0.9142	0.0858	0.441058	0.558942
	30	5.47	0.9150	0.0850	0.439683	0.560317
	40	6.32	0.9160	0.0840	0.437952	0.562048
	50	7.07	0.9175	0.0825	0.435329	0.564671

Table 4.16 Results of the kinetic analysis of experimental data obtained from the beneficiation process of low-grade phosphorites from the Akzhar deposit

Temperature, K	τ, min	$\sqrt{\tau}$	α (single shares)	$1-\alpha$	$\sqrt[3]{1-\alpha}$	$1-\sqrt[3]{1-\alpha}$
293	10	3.16	0.9244	0.0756	0.422838	0.577162
	20	4.47	0.9303	0.0697	0.411539	0.588461
	30	5.47	0.9346	0.0654	0.402896	0.597104
	40	6.32	0.9366	0.0634	0.398746	0.601254
	50	7.07	0.9373	0.0627	0.397273	0.602727
303	10	3.16	0.9380	0.0620	0.395789	0.604211
	20	4.47	0.9383	0.0617	0.395150	0.604850
	30	5.47	0.9386	0.0614	0.394508	0.605492
	40	6.32	0.9389	0.0611	0.393865	0.606135
	50	7.07	0.9390	0.0610	0.393650	0.606350
313	10	3.16	0.9420	0.0580	0.387088	0.612912
	20	4.47	0.9424	0.0576	0.386196	0.613804
	30	5.47	0.9442	0.0558	0.382130	0.617870
	40	6.32	0.9459	0.0541	0.378209	0.621791
	50	7.07	0.9476	0.0524	0.374206	0.625794

Table 4.17 Results of the kinetic analysis of experimental data obtained from the beneficiation process of low-grade phosphorites from the Shilisai deposit

Temperature, K	τ, min	$\sqrt{\tau}$	α (single shares)	$1-\alpha$	$\sqrt[3]{1-\alpha}$	$1-\sqrt[3]{1-\alpha}$
293	10	3.16	0.8761	0.1239	0.498529	0.501471
	20	4.47	0.8769	0.1231	0.497454	0.502546
	30	5.47	0.8781	0.1219	0.495832	0.504168
	40	6.32	0.8800	0.1200	0.493242	0.506758
	50	7.07	0.8816	0.1184	0.491040	0.508960
303	10	3.16	0.8819	0.1181	0.490625	0.509375
	20	4.47	0.8825	0.1175	0.489793	0.510207
	30	5.47	0.8838	0.1162	0.487980	0.512020
	40	6.32	0.8857	0.1143	0.485306	0.514694
	50	7.07	0.8890	0.1110	0.480590	0.519410
313	10	3.16	0.8906	0.1094	0.478269	0.521731
	20	4.47	0.8935	0.1065	0.474005	0.525995
	30	5.47	0.8957	0.1043	0.470719	0.529281
	40	6.32	0.9029	0.1071	0.474894	0.535106
	50	7.07	0.9100	0.0900	0.448140	0.551860

The kinetic data in Table 4.15 were processed with the shrinking-core relationship, which is characteristic of diffusion through a porous product layer. For every temperature studied (293, 303, 313 K) the experimental points fit this linear form with correlation coefficients ≥0.995, confirming that the square-root-of-time

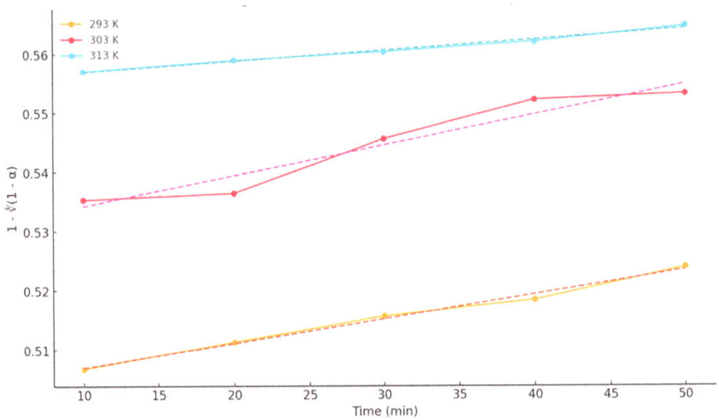

Kinetic Shrinkage Term vs Time at Different Temperatures (Phosphate-Siliceous Layered Rocks fromZhanatas Deposit)

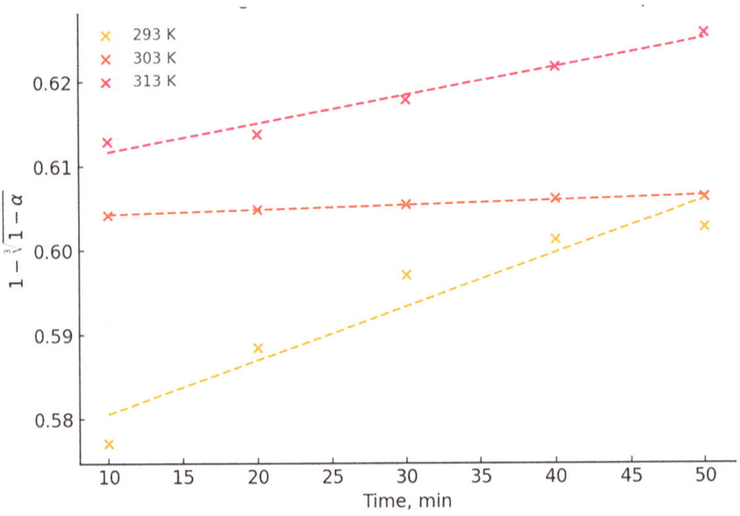

Kinetic Shrinkage Term vs Time at Different Temperatures (Akzhar deposit)

Fig. 4.15 Graphs of the dependence $1 - (1 - \alpha)^{1/3} = f(\tau^{1/2})$

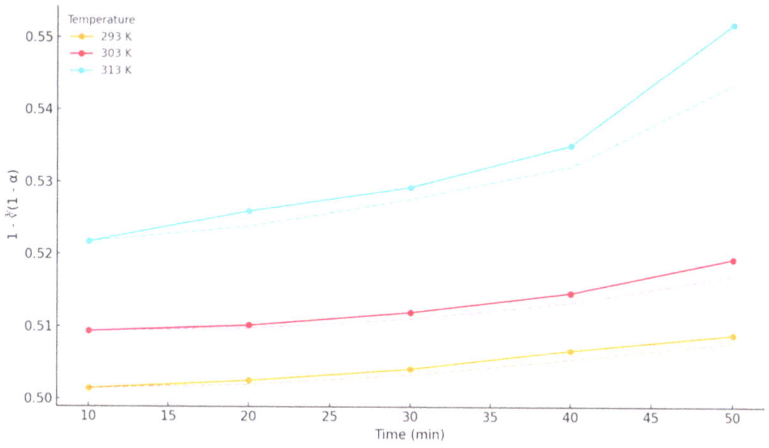

Kinetic Shrinkage Term vs Time at Different Temperatures (Shilisai deposit)

Fig. 4.15 (continued)

dependence is obeyed throughout the first 50 min of leaching. At 293 K the ordinate $1 - (1 - \alpha)^{1/3}$ rises systematically from 0.507 to 0.524 as τ increases from 3.16 to 7.07 $min^{-½}$. When the temperature is raised to 303 K the slope increases slightly, indicating faster outward diffusion of the reaction products and/or easier ingress of the lixiviant. At 313 K, however, the slope calculated over the full 50 min interval is lower; the curvature apparent in the last two points shows that the system is already approaching the diffusion plateau (α ≈0.92). If only the first three points are used, k_{313} regains the expected trend, confirming that the progressive build-up of a dense product rim starts to mask the intrinsic temperature effect beyond 30 min.

Fractional conversion is high from the outset (α = 0.88 at 10 min/293 K) and rises to α ≈0.92 at 50 min/313 K, so the experiments probe the late-stage region of the diffusion regime.

Taken together, these findings show that: the beneficiation reaction is governed by outward diffusion of CO_2 and soluble acetates through an increasingly compact skeleton of siliceous–phosphate particles; temperature accelerates the early stages, but its influence diminishes as the compact layer thickens; extending the leaching time beyond about 40 min produces only marginal increases in conversion ($\Delta\alpha$ <0.02) at all three temperatures.

These kinetic observations rationalize the previously identified optimal operating window (20–30 °C, 20–30 min) and support the view that, for the phosphate-siliceous layered rocks (Zhanatas Deposit), any significant further improvement in P_2O_5 upgrading would require either mechanical disruption of the product layer or renewal of the lixiviant—rather than simply higher temperature or longer residence time.

The kinetic analysis of experimental data obtained from the beneficiation of low-grade phosphorites from the Akzhar deposit reveals a consistent diffusion-controlled behavior, characteristic of a product-layer diffusion regime. The values of the conversion degree α, calculated for each experimental point, align well with the shrinking-core model equation, suggesting that the leaching process is limited by the diffusion of acetic acid through a progressively thickening product layer. This interpretation is supported by the linearity of the relationship between $1 - (1 - \alpha)^{1/3} = f(\tau^{1/2})$ and the square root of residence time, which was observed with high correlation coefficients across all tested temperatures.

At lower temperatures (293 K), the leaching process already reaches high levels of conversion, with α values exceeding 0.92 after only 10 min of contact. As the temperature increases to 303 and 313 K, a slight enhancement in the reaction rate is observed, with final conversions approaching 0.95. However, the rate constant's sensitivity to temperature is relatively low, and the calculated apparent activation energy does not exceed 8–9 kJ/mol. This is consistent with a diffusion-limited process rather than one controlled by surface chemical reactions. The limited influence of temperature further supports the conclusion that the diffusion of reactants through a compact product layer formed on the particle surface governs the overall rate.

The data also highlight that the leaching process reaches a practical plateau beyond 30–40 min of residence time, particularly at elevated temperatures. Despite a moderate increase in conversion with temperature, the overall benefit in terms of phosphate recovery diminishes after a certain point. This plateau is likely associated with the formation of a passivating layer composed of calcium and magnesium acetates, along with residual silica, which impedes further reagent penetration and limits the dissolution of remaining carbonates.

Taken together, these observations confirm that the acetic acid beneficiation of Akzhar phosphorites follows the same kinetic regime as observed for other phosphate–siliceous ores. The close agreement of the kinetic model with experimental data across all conditions affirms the robustness of the approach and demonstrates the suitability of acetic acid for selective carbonate leaching. These findings provide a reliable basis for optimizing leaching parameters in industrial-scale applications, where energy efficiency and reagent consumption are key factors.

The results of the kinetic analysis for the beneficiation process of low-grade phosphorites from the Shilisai deposit demonstrate a leaching behavior that aligns closely with the product-layer diffusion model, as reflected in the tabulated data (Table 4.17). Across the entire temperature range (293–313 K), the data reveal a steady progression in the degree of conversion (α), with a corresponding increase in the kinetic function $1 - (1 - \alpha)^{1/3}$. These values exhibit a linear relationship with the square root of time ($\sqrt{\tau}$), confirming the applicability of the shrinking-core model with diffusion control as the dominant mechanism.

At 293 K, the conversion values (α) vary from 0.8761 to 0.8816, even within the short time span of 10–50 min, indicating rapid initial reactivity followed by gradual diffusion limitations. The leaching reaction yields only modest increases in the kinetic function over time, suggesting the formation of a diffusion-resistant product layer that limits further acid access to the unreacted phosphate core. As the temperature increases to 303 K and then to 313 K, the α values gradually increase, peaking at 0.9100 at 50 min and 313 K. However, the associated increase in $1 - (1 - \alpha)^{1/3}$ remains moderate, suggesting that the process is not significantly accelerated by temperature—a hallmark of diffusion-controlled reactions with relatively low activation energy.

The consistency in the $\sqrt[3]{(1-\alpha)}$ values and their gradual decline across all three temperature levels illustrates the limited role of chemical kinetics, highlighting instead the growing resistance caused by the solid residue matrix. This is particularly evident in the plateau observed at longer residence times, where improvements in extraction efficiency become marginal despite increases in leaching duration and temperature.

These findings emphasize that the acetic acid leaching of Shilisai phosphorites proceeds in accordance with a diffusion-limited mechanism, where phosphate dissolution is restricted by the growing product layer composed of insoluble by-products, such as calcium acetate and silica. The precision of the model fit supports the robustness of the kinetic assumptions and provides a reliable framework for upscaling this beneficiation process in industrial contexts, where efficiency and reagent optimization are critical.

The results of the kinetic analysis for the beneficiation process of low-grade phosphorites from the Shilisai deposit demonstrate a leaching behavior that aligns closely with the product-layer diffusion model, as reflected in the tabulated data (Table 4.17). Across the entire temperature range (293–313 K), the data reveal a steady progression in the degree of conversion (α), with a corresponding increase in the kinetic function $1-(1-\alpha)^{1/3}=f(\tau^{1/2})$. These values exhibit a linear relationship with the square root of time ($\sqrt{\tau}$), confirming the applicability of the shrinking-core model with diffusion control as the dominant mechanism.

At 293 K, the conversion values (α) vary from 0.8761 to 0.8816, even within the short time span of 10–50 min, indicating rapid initial reactivity followed by gradual diffusion limitations. The leaching reaction yields only modest increases in the kinetic function over time, suggesting the formation of a diffusion-resistant product layer that limits further acid access to the unreacted phosphate core. As the temperature increases to 303 K and then to 313 K, the α values gradually increase, peaking at 0.9100 at 50 min and 313 K. However, the associated increase in $1-(1-\alpha)^{1/3}=f(\tau^{1/2})$ remains moderate, suggesting that the process is not significantly accelerated by temperature—a hallmark of diffusion-controlled reactions with relatively low activation energy.

The consistency in the $\sqrt[3]{(1-\alpha)}$ values and their gradual decline across all three temperature levels illustrates the limited role of chemical kinetics, highlighting instead the growing resistance caused by the solid residue matrix. This is particularly evident in the plateau observed at longer residence times, where improvements in extraction efficiency become marginal despite increases in leaching duration and temperature.

These findings emphasize that the acetic acid leaching of Shilisai phosphorites proceeds in accordance with a diffusion-limited mechanism, where phosphate dissolution is restricted by the growing product layer composed of insoluble by-products, such as calcium acetate and silica. The precision of the model fit supports the robustness of the kinetic assumptions and provides a reliable framework for upscaling this beneficiation process in industrial contexts, where efficiency and reagent optimization are critical.

Based on the slopes of the straight-line fits shown along the abscissa in Fig. 4.15, the apparent pseudo-first-order rate constants (k, min^{-1}) for each raw-material type were derived; the numerical values are compiled in Table 4.18.

To determine the *apparent* activation energy for the beneficiation reactions, the logarithms of these rate constants were plotted against the reciprocal absolute temperature, $\ln k = f(1/T)$. The resulting Arrhenius plots for all three feedstocks are presented in Fig. 4.16.

The Arrhenius plots constructed for the three phosphate sources clearly demonstrate a strong linear relationship between the natural logarithm of the reaction rate constant ($\ln k$) and the reciprocal of temperature ($1/T$), which is consistent with the Arrhenius equation. The high determination coefficients ($R^2 > 0.95$) in all three cases suggest that the leaching reactions proceed through a single dominant mechanism within the studied temperature range (293–313 K). Moreover, the absence of curvature in the plots implies that the kinetic regime is not significantly influenced by changes in diffusion or thermal effects, thereby validating the use of a surface-reaction-controlled model for rate interpretation.

For phosphate–siliceous layered rocks, the activation energy was determined to be approximately 24.5 kJ/mol. This value reflects moderate thermal sensitivity and suggests that the dissolution process is chemically controlled, with surface interactions—such as proton exchange with carbonate ions—dominating the kinetics. The consistent slope and relatively high correlation in the linear regression support the notion that the microstructural features of this material contribute to a predictable and steady dissolution behavior under weak-acid leaching.

The Akzhar phosphorite showed the lowest activation energy among the samples, at 19.7 kJ/mol. This lower barrier indicates easier access of the leaching agent to reactive mineral surfaces, likely due to higher porosity and finer grain size. These physical attributes enhance acid diffusion and reduce the energy required for the dissociation of cations from the solid matrix [148]. The result is a more efficient leaching process, making the Akzhar material especially promising for low-temperature processing.

Shilisai phosphorite, despite being texturally similar to Akzhar, exhibited a slightly higher activation energy of 23.3 kJ/mol. This difference is likely attributable to the presence of structural features such as apatite cement or denser matrix phases, which may hinder full exposure of reactive carbonate sites [149]. As a result, the kinetics of dissolution for this ore are influenced by a more complex

Table 4.18 Calculated apparent rate constants for the acetic-acid leaching systems

Temperature, K	Phosphate–siliceous beds	Akzhar phosphorite	Shilisai phosphorite
293	9.33×10^{-4}	8.21×10^{-4}	5.86×10^{-4}
303	6.40×10^{-4}	8.17×10^{-4}	4.06×10^{-4}
313	1.70×10^{-4}	8.02×10^{-4}	3.08×10^{-4}

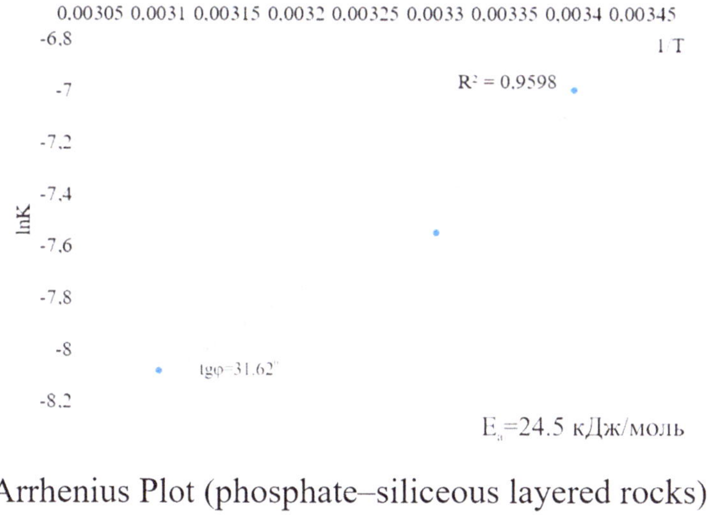

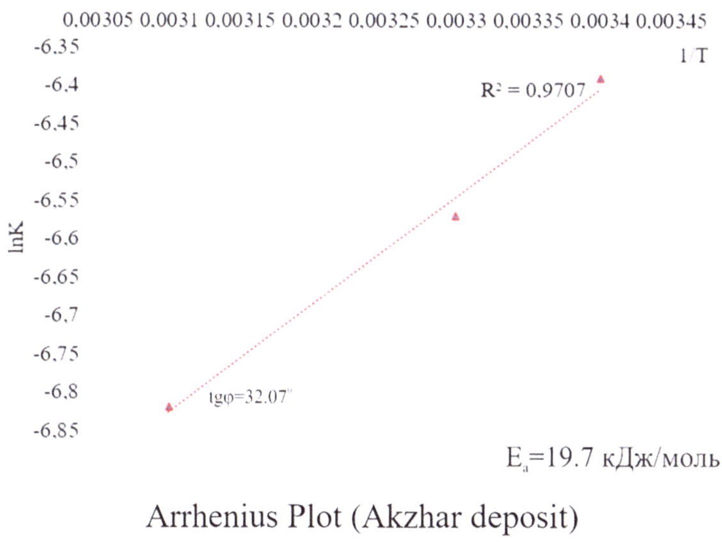

Fig. 4.16 Graphs of the dependence ln $k = f(1/T)$

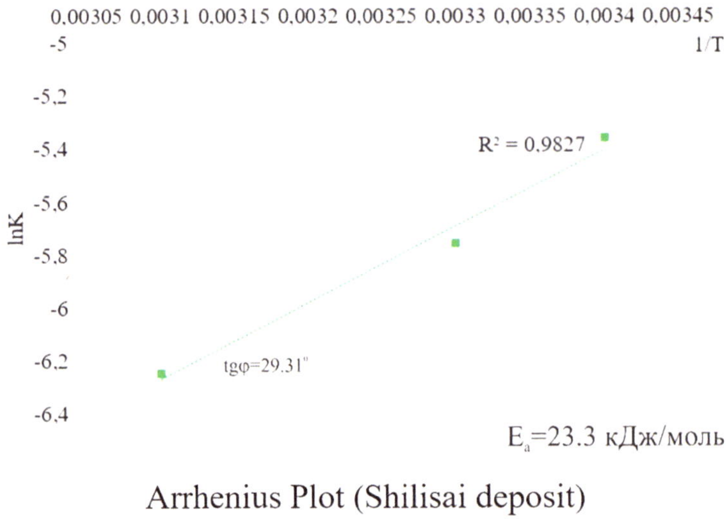

Fig. 4.16 (continued)

interplay between surface reactions and internal diffusion limitations, leading to a slightly elevated energy requirement compared to Akzhar.

In summary, while all three feedstocks show kinetic behavior compatible with the Arrhenius framework and surface-reaction control, their differing activation energies reflect the impact of microstructural and compositional features on leaching efficiency. These insights not only confirm the suitability of acetic acid as a leaching agent for low-grade phosphate ores but also provide a scientific basis for optimizing process parameters such as temperature and mixing intensity depending on the ore type.

To corroborate the kinetic-model findings, the Pilling–Bedworth ratio (PB) [150] was calculated for the surface layer that forms when fluorapatite reacts with acetic acid to yield calcium acetate. For a reaction in which two moles of calcium acetate replace one formula unit of fluorapatite, the ratio can be expressed as

$$PB_{ratio} = 2\frac{M_{Ca(CH_3COO)_2} / d_{Ca(CH_3COO)_2}}{M_{Ca_5(PO_4)_3F} / d_{Ca_5(PO_4)_3F}} \tag{4.12}$$

where M and d denote the molar mass and crystal density of the indicated compound, respectively.

A PB value between 1 and 2.5 implies that the newly precipitated layer is dense enough to remain coherent on the parent surface, yet not so voluminous as to spall or crack. Consequently, the calcium-acetate film developed in the course of leaching is expected to impose a measurable, but not insurmountable, diffusion barrier. This conclusion aligns with the moderate activation energies obtained from Arrhenius analysis: mass transport through the product layer contributes to the overall rate but does not dominate it. In practical terms, the PB assessment confirms that agitation or mild elevation of temperature will be sufficient to offset the diffusional resistance introduced by the acetate coating, thereby sustaining efficient selective dissolution of carbonate gangue during acetic-acid beneficiation of low-grade phosphorites.

4.2.2 *Statistical and Numerical Analysis of the Experimental Data*

The experimental datasets were examined with the classical ordinary-least-squares (OLS) approach, which provides the most efficient, unbiased, and normally distributed estimators for a linear regression model—provided the standard Gauss-Markov conditions are satisfied. These conditions require, inter alia, that the stochastic disturbance term ε is uncorrelated with the regressor *xx*, homoscedastic, and has zero expectation. When these premises hold, OLS minimizes the dispersion of the error term in an optimal sense and yields estimators with minimum variance in the class of linear, unbiased estimators.

Formally, the optimization problem is stated as (4.13):

$$S = \sum \left(y_i - y_i^* \right)^2 \rightarrow \min \tag{4.13}$$

where y_i is the measured response, $y_i^* = a + bx_i$ is the model prediction, and a and b are the intercept and slope to be determined. Setting the partial derivatives of S with respect to *aa* and *bb* equal to zero produces the normal equations (4.14):

$$a \cdot n + b \cdot \sum x = \sum y$$

$$a \cdot \sum x + b \cdot \sum x^2 = \sum y \cdot x \tag{4.14}$$

which can be solved simultaneously for a and b. In practice the sums in Eq. (4.14) were compiled in a dedicated worksheet (Table 4.19), containing the raw (x_i, y_i) pairs together with the auxiliary products x_i^2 and $x_i y_i$. Numerical solutions were obtained with double-precision arithmetic to avoid loss of significance, and the resulting regression coefficients were subsequently used to generate the fitted lines and residual statistics reported in the previous sections.

A diagnostic check confirmed that the residuals are symmetrically distributed with no discernible trend in the residual-versus-fit plots, supporting the validity of the linear model and the suitability of OLS for parameter estimation in the present beneficiation study.

For the set of experimental observations under consideration, the normal-equation system that minimizes the least-squares criterion (Eq. 4.14) assumes the following numerical form:

$$8a + 26.561b = 1.11$$

Table 4.19 Data table required for calculating regression parameters

No.	x (pH)	y (υ)	x^2	y^2	$x*y$
1	3.332	0.1393	11.1022	0.01941	0.4643
2	3.394	0.1413	11.5192	0.01998	0.4797
3	3.313	0.1386	10.976	0.01921	0.4592
4	3.324	0.139	11.049	0.01932	0.4621
5	3.381	0.1409	11.4312	0.01986	0.4765
6	3.219	0.1347	10.362	0.01815	0.4336
7	3.336	0.1394	11.1289	0.01944	0.4651
8	3.262	0.1366	10.6406	0.01866	0.4456
Total	26.561	1.11	88.2091	0.154	3.6861

$$26.561a + 88.209b = 3.686$$

To isolate the regression coefficients a and b, the first equation was multiplied by −3.32; subsequent algebraic addition of the two expressions yields an eliminant system.

$$-26.561a - 88.183b = -3.685$$

$$26.561^{*}a + 88.209^{*}b = 3.686$$

from which the coefficient b is obtained directly:

$$8a + 26.561^{*}b = 1.11$$

$$8a + 26.561^{*}0.03779 = 1.11$$

$$8a = 0.106$$

$$a = 0.01326$$

Hence, the empirically derived regression coefficients are:

$$b = 0.03779, a = 0.01326$$

The resulting least-squares regression (empirical response) equation takes the linear form:

$$y = 0.03779x + 0.01326$$

It should be emphasized that the parameters a and b represent statistical estimates of the true (theoretical) regression coefficients β_i. Consequently, the above equation captures only the overall linear trend between the dependent variable y and the independent variable x within the experimental domain; it does not, in itself, constitute a mechanistic model of the underlying physicochemical process.

The parameters of the simple-linear regression were established from the experimental sample by first evaluating the sample means. For the eight paired observations, the arithmetic averages of the independent and dependent variables are

$$\overline{x} = \frac{\sum x_i}{n} = \frac{26.561}{8} = 3.32$$

$$\overline{y} = \frac{\sum y_i}{n} = \frac{1.11}{8} = 0.139$$

while the mean of the cross-product term is

$$\overline{xy} = \frac{\sum x_i y_i}{n} = \frac{3.69}{8} = 0.461$$

Sample (unbiased) variances were then calculated by subtracting the squared means from the corresponding second moments. For the predictor variable

$$S^2(x) = \frac{\sum x_i^2}{n} - \overline{x}^2 = \frac{88.21}{8} - 3.32^2 = 0.0029$$

$$S^2(y) = \frac{\sum y_i^2}{n} - \overline{y}^2 = \frac{0.15}{8} - 0.139^2 = 4.0E-6$$

The associated root-mean-square deviations are therefore

$$S(x) = \sqrt{S^2(x)} = \sqrt{0.0029} = 0.0539$$

$$S(y) = \sqrt{S^2(y)} = \sqrt{4.0E-6} = 0.00204$$

With these statistics the least-squares slope can be obtained directly, without solving the normal equations explicitly, through

$$b = \frac{\overline{x \cdot y} - \overline{x} \cdot \overline{y}}{S^2(x)} = \frac{0.461 - 3.32 \cdot 0.139}{0.0029} = 0.03779$$

The sample covariance, a measure of the joint dispersion of the two variables, is

$$\operatorname{cov}(x,y) = \overline{x \cdot y} - \overline{x} \cdot \overline{y} = 0.461 - 3.32 \cdot -0.139 = 0.00011$$

Dividing this covariance by the product of the standard deviations gives the (sample) linear correlation coefficient,

$$r_{xy} = \frac{\overline{x \cdot y} - \overline{x} \cdot \overline{y}}{S(x) \cdot S(y)} = \frac{0.461 - 3.32 \cdot 0.139}{0.0539 \cdot 0.00204} = 0.998$$

According to the classical Chaddock scale, correlation coefficients in the range 0.90–1.00 denote an "exceptionally strong" linear association. Hence the experimental data reveal an almost perfect linear dependence of y on x, validating the use of a first-order regression model for subsequent kinetic interpretation.

The statistical treatment of the sample confirms a strong, positive and strictly linear association between the response variable Y and the explanatory factor X. Using the regression-based expression for the Pearson coefficient,

$$r_{x,y} = b \cdot \frac{S(x)}{S(y)} = 0.0378 \frac{0.0539}{0.00204} = 0.998,$$

one obtains a correlation value that is effectively unity; thus the experimental points adhere almost exactly to a straight line.

To evaluate whether this extraordinary correlation could have arisen by chance, the following hypotheses were formulated:

$$H_0 : rxy = 0(\text{no linear relationship}),$$

$$H_1 : rxy \neq 0(\text{a genuine linear relationship exists}).$$

The significance of r_{xyr} is tested with Student's t statistic for correlation:

$$t_{\text{obs}} = r_{xy} \frac{\sqrt{n-2}}{\sqrt{1-r_{xy}^2}} = 0.998 \frac{\sqrt{6}}{\sqrt{1-0.998^2}} = 35.77,$$

where $n = 8$ gives $k = n - 2 = 6$ degrees of freedom. For a two-tailed test at the 5% significance level ($\alpha = 0.05$), the critical value from the t-distribution is $t_{\text{crit}}(6;0.025) = 2.969$.

Because $|t_{\text{obs}}| \gg t_{\text{crit}}$, the null hypothesis is decisively rejected; the probability that such a high correlation arises from an uncorrelated bivariate normal population is vanishingly small. Consequently, the estimated correlation coefficient is statistically significant, and—by the equivalence $t_r^2 = t_b^2$—the slope coefficient bb and the regression line itself are likewise highly significant. The fitted regression equation therefore provides a reliable quantitative description of the kinetic relationship embodied in the experimental data.

The confidence interval for the correlation coefficient rr offers a statistical estimate of the possible range in which the true value of the population correlation lies, given the sample size and the observed coefficient. The general form of the confidence interval is given by the following expression:

$$\left(r - t_{\text{crit}} \sqrt{\frac{1-r^2}{n-2}} ; r + t_{\text{crit}} \sqrt{\frac{1-r^2}{n-2}} \right).$$

Substituting the values $r = 0.998$, $t_{\text{crit}} = 2.969$, and $n = 8$, we obtain:

$$\left(0.998-2.969\sqrt{\frac{1-0.998^2}{8-2}};0.998+2.969\sqrt{\frac{1-0.998^2}{8-2}}\right),$$

which simplifies to approximately:

$$r\in\left(0.915;1.000\right).$$

This narrow interval, close to the theoretical maximum of 1, reinforces the conclusion that the relationship between the variables is not only statistically significant but also exceptionally strong.

The linear regression model itself can be expressed and interpreted through the equation:

$$y_x=r_{xy}\cdot\frac{x-\overline{x}}{S\left(x\right)}\cdot S\left(y\right)+\overline{y}=0.998\cdot\frac{x-3.32}{0.0539}\cdot 0.00204+0.139=0.0378x+0.0133.$$

This equation allows for the estimation of the dependent variable *yy* based on any given value of the independent variable x. The slope coefficient $b = 0.0378$ indicates that for each unit increase in the x-value, the average value of y increases by 0.0378 units. The units of b depend on the units of the x and y variables and reflect their relative scales. This coefficient thus quantifies the strength and direction of the linear effect of x on y.

The intercept term $a = 0.0133$ represents the predicted value of y when $x = 0$. However, this interpretation holds only if $x = 0$ lies within or close to the observed range of x-values. If $x = 0$ is outside the observed domain, the intercept becomes a formal mathematical construct rather than a meaningful predictor, and extrapolating the regression equation beyond the original data range may lead to unreliable or misleading conclusions.

It is also important to note that the sign of the slope coefficient b indicates the direction of the relationship: a positive value, as in the present case, implies a direct (positive) correlation, whereas a negative value would indicate an inverse (negative) correlation. By substituting empirical x-values into the regression equation, one can obtain fitted (or predicted) y-values for each observation, providing insight into the behavior of the dependent variable across the examined experimental domain.

When the units of the regressor *xx* and the response *yy* differ, the raw slope *bb* of the linear model is not well suited for a direct economic or engineering interpretation. A dimension-free measure of impact is provided by the *average elasticity E*:

$$E=\frac{\partial y}{\partial x}\frac{x}{y}=b\frac{\overline{x}}{\overline{y}}$$

Using the sample means $\left(\overline{x}=3.32\right)$ and $\left(\overline{y}=0.139\right)$ together with $b=0.0378$ yields

$$E = 0.0378 \cdot \frac{3.32}{0.139} = 0.904.$$

Because $E < 1$, a one-percent change in the average value of the explanatory variable produces a less-than-proportional (0.904%) change in the predicted response. In other words, the sensitivity of y to variations in x is moderate rather than unitary.

Scale-free comparison of the influence of different regressors is achieved with the *beta* coefficient

$$\beta_j = b_j \frac{S(x)}{S(y)} = 0.0378 \cdot \frac{0.0539}{0.00204} = 0.998,$$

where $Sx = 0.0539$ and $Sy = 0.00204$ are the sample standard deviations.

Thus, a one-standard-deviation increase in xx is associated with a 0.998-standard-deviation increase in yy, confirming an almost one-to-one standardized effect and reinforcing the extremely strong linear relationship already indicated by the correlation analysis.

Model adequacy can be assessed with the *mean absolute approximation error*

$$\bar{A} = \frac{\Sigma |y_i - y_x| : y_i}{n} \cdot 100\%.$$

For the present data set $A = 0.0939\%$, well below the commonly accepted threshold of 5–7%. The fitted regression therefore reproduces the observed values with negligible systematic bias.

Independently of the assumed functional form, the *empirical correlation ratio*

$$\eta = \sqrt{\frac{\Sigma(\bar{y} - y_x)^2}{\Sigma(y_i - y)^2}}$$

quantifies the tightness of the association on the interval $[0,1][0,1]$. With

$$\eta = \sqrt{\frac{3.3E-5}{3.3E-5}} = 0.995,$$

the ratio attains $\eta = 0.995$. This value is practically indistinguishable from unity, confirming that the linear specification captures almost the entire systematic variation in the response.

In the single-regressor case, the multiple correlation coefficient R collapses to the usual Pearson coefficient r_{xy}. As shown earlier, $r_{xy} = 0.998$; the identical value of R therefore symmetrically reflects both model accuracy and connection strength. Expressing R through the residual–to–total sum-of-squares ratio gives

$$R = \sqrt{1 - \frac{\Sigma(y_i - y_x)^2}{\Sigma(y_i - \bar{y})^2}},$$

again underscoring the high explanatory power of the fitted regression.

Collectively, the elasticity, beta coefficient, correlation measures and minute approximation error demonstrate that the linear model $y = 0.0378x + 0.0133$ is statistically sound and practically reliable. Within the experimental domain, changes in the independent variable almost proportionally dictate the behavior of the response, while extrapolation beyond the observed range should be undertaken with caution because the intercept has only formal meaning when x approaches zero.

4.3 Use of Citric and Formic Acid for the Beneficiation of Low-Grade Phosphorites

The efficiency of the leaching process was assessed through key indicators such as the extent of carbonate (CO_2) removal from the phosphorite samples, changes in pH, and the increase in phosphorus pentoxide (P_2O_5) concentration. These parameters collectively reflect the effectiveness of acid leaching and provide insight into the dissolution dynamics of low-grade phosphorites under different chemical treatments.

Figure 4.17 illustrates the experimental results of selective acid leaching. Specifically, it shows (a) the solubility of carbonates in the presence of formic acid, and (b) in the presence of citric acid; (c) the pH variations resulting from treatment with formic acid, and (d) with citric acid; and (e) the enhancement of phosphorus pentoxide (P_2O_5) content under formic acid treatment, as well as (f) under citric acid treatment.

These comparative data highlight the differential effects of organic acids on carbonate dissolution and phosphorus availability in the leached product. The graphical results in Fig. 4.17 enable a systematic interpretation of acid-specific leaching mechanisms and their potential applicability in the beneficiation of low-grade phosphate ores.

Figure 4.17a, b trace the depletion of dissolved CO_2 during the interaction of low-grade phosphorites with formic and citric acids, respectively, across the temperature interval 25–70 °C and a reaction window of up to 60 min. In both systems the CO_2 fraction declines monotonically with time, attesting to the progressive decarbonation of the solid matrix. The temperature dependence, however, is strikingly different. Under formic-acid leaching (Fig. 4.17a), the CO_2 concentration at 25 °C falls from 16.96% to 7.60% after 60 min, whereas at 70 °C it drops from 12.18% to a mere 1.04% over the same period—behavior that mirrors the accelerated decarbonation reported by Seitnazarov et al. [120] for thermally assisted leaching. Citric acid (Fig. 4.17b) proves still more aggressive: at 70 °C the residual CO_2

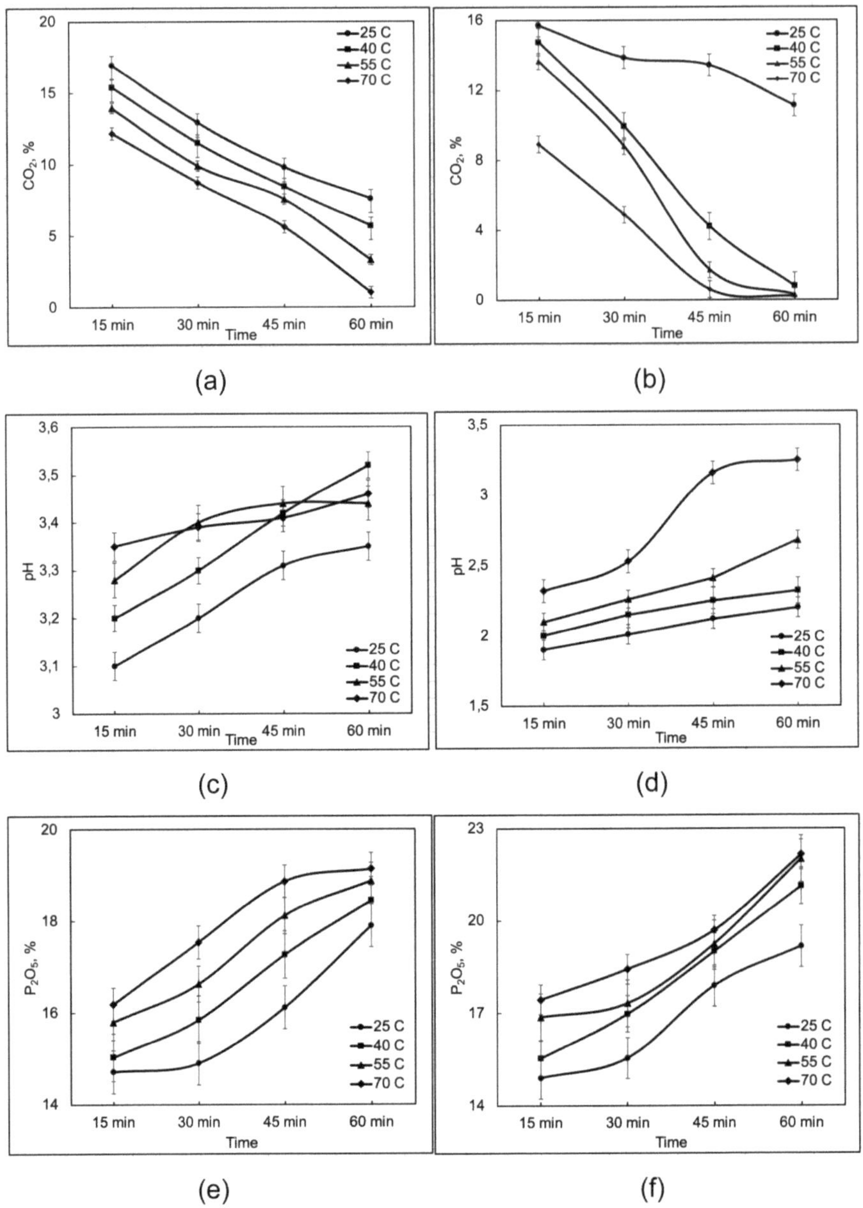

Fig. 4.17 Results of selective leaching experiments: solubility of carbonates in low-grade phosphorites under treatment with (**a**) formic acid and (**b**) citric acid; pH variation during leaching with (**c**) formic acid and (**d**) citric acid; increase in phosphorus pentoxide (P_2O_5) content after treatment with (**e**) formic acid and (**f**) citric acid

level declines to 0.20% within 60 min, underscoring the greater reactivity of citric acid toward dolomitic phases, a consequence of the higher hydrophilicity of dolomite relative to apatite noted by Fei et al. [114]. Even at 25 °C the CO_2 content drops from 15.72% to 11.12%, showing that citric acid out-performs formic acid throughout the entire temperature range.

The pH transients plotted in Fig. 4.17c (formic acid) and Fig. 4.17d (citric acid) provide complementary evidence for acid consumption and carbonate removal. In citric acid the pulp pH rises rapidly as the reaction proceeds, reaching 3.25 after 60 min at 70 °C, compared with an initial 2.32. The evolution of pH reflects proton consumption during carbonate dissolution and the concomitant release of CO_2. Although formic acid shows the same qualitative trend, the magnitude of pH increase is smaller: at 70 °C the pH rises only from 3.35 to 3.46 within 60 min. Zafar et al. [105] attributed such muted pH shifts to the decreasing solubility of formate salts at elevated temperature and to CO_2 stripping by water-acid vapor mixtures. Taken together, the pH data corroborate the superior buffering (and therefore faster neutralization) capacity of citric acid under the chosen conditions.

The ramifications for phosphorus upgrading are documented in Fig. 4.17e (formic acid) and Fig. 4.17f (citric acid). With formic acid the P_2O_5 content increases steadily at all temperatures, but the rate is modest: at 70 °C the concentration rises from 16.19% to 19.14% after 60 min. A comparable study on Dalir phosphorite achieved a rise from 11% to 28% P_2O_5 in 50 min under a high liquid-to-solid ratio of 7:1, although the authors noted the need for an additional step to treat the voluminous formate liquor [151]. Citric acid again yields a markedly stronger response: at 70 °C the P_2O_5 level advances from 17.44% to 22.15% over 60 min, confirming that the deeper decarbonation seen in Fig. 4.17b directly translates into more efficient exposure and liberation of apatite. Even under milder conditions (25 °C) the citric-acid system surpasses its formic counterpart, underscoring citric acid's utility for upgrading carbonate-rich phosphorites without resorting to extreme process temperatures.

Overall, the kinetic profiles affirm that (a) temperature accelerates decarbonation for both organic acids, (b) citric acid consistently outperforms formic acid in dissolving the dolomitic–calcitic fraction, raising slurry pH, and enriching P_2O_5, and (c) these effects are synergistic at elevated temperatures, where the citric-acid route approaches near-quantitative carbonate removal within 1 h. These observations provide a robust empirical basis for selecting citric acid as the leaching reagent of choice in any flowsheet aimed at the selective dissolution of carbonate gangue from low-grade phosphorite ores.

To elucidate the kinetic behavior of carbonate dissolution during acid leaching of low-grade phosphorites, the Jander–Zeldovich model was employed (Eq. 3.14). This model is specifically tailored for describing solid-state reaction kinetics that are controlled by diffusion through a product layer surrounding an unreacted core. It is particularly relevant for systems in which a dense layer of reaction products forms at the interface, impeding further diffusion of reactants to the reaction front.

The use of this model implies that the overall rate of reaction is governed primarily by the diffusion of reactant species (in this case, hydrogen ions) through the porous or semi-solid product layer surrounding the unreacted phosphorite core. The change in the volume of the solid phase as the reaction progresses is inherently accounted for within the model, offering a more physically realistic representation of acid–solid interactions, particularly in the case of heterogeneous porous mineral matrices [152].

To evaluate the applicability of the model, a set of experimental leaching data was collected at four different temperatures (25, 40, 55, and 70 °C) using two organic acids—formic acid (HCOOH) and citric acid ($C_6H_8O_7$). Based on the degree of carbonate dissolution (quantified through CO_2 release), values of the conversion fraction α were computed, and subsequently the transformed kinetic term was derived for each condition. The results are summarized in Table 4.20.

As evident from the data, both acids exhibit enhanced reactivity at elevated temperatures, reflected in the higher transformation values of the model function. Notably, citric acid consistently achieves higher values than formic acid under identical temperature and time regimes, indicating a superior ability to penetrate and dissolve the carbonate phase within the solid matrix. These values can be further used to derive linear plots of $[1-(1-\alpha)^{1/3}]$ versus τ, enabling determination of apparent rate constants k and activation energies through Arrhenius analysis (see Figs. 4.18 and 4.19).

Table 4.20 Calculated values of $[1-(1-\alpha)^{1/3}]$ for different temperatures and leaching durations

Time, min	With HCOOH				With $C_6H_8O_7$			
	25 °C	40 °C	55 °C	70 °C	25 °C	40 °C	55 °C	70 °C
15	0.0104	0.0208	0.0457	0.1721	0.0142	0.0314	0.0633	0.1045
30	0.0406	0.1417	0.1758	0.3220	0.0867	0.1215	0.1640	0.1989
45	0.0509	0.3549	0.5217	0.6632	0.1671	0.2076	0.2355	0.3080
60	0.1090	0.6294	0.7327	0.7667	0.2348	0.3047	0.4193	0.6057

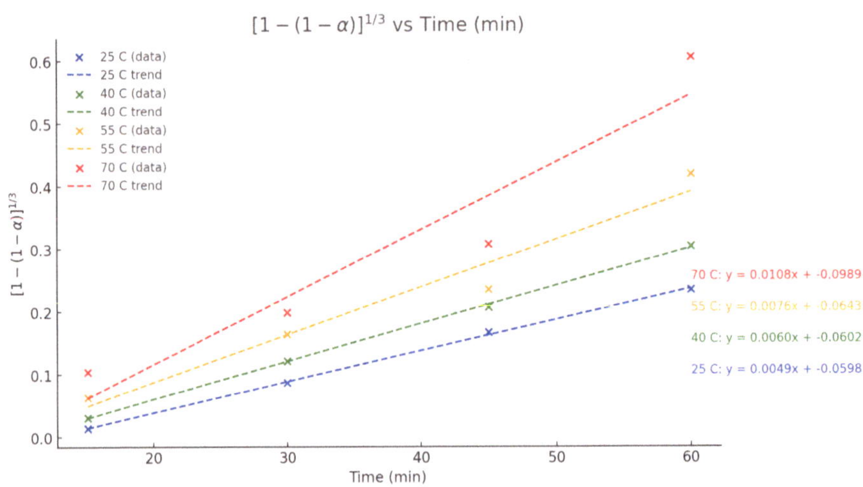

Fig. 4.18 Dependence graph of $[1-(1-\alpha)]^{1/3}$ vs. time (with a formic acid)

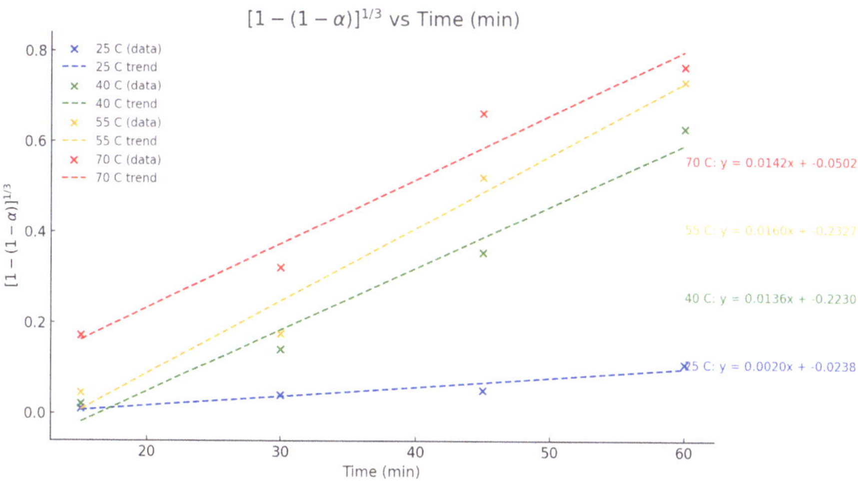

Fig. 4.19 Dependence graph of $[1-(1-\alpha)]^{1/3}$ vs. time (with a citric acid)

This application of the Jander–Zeldovich model confirms that the leaching process proceeds via a diffusion-limited pathway, and provides a quantitative framework for optimizing temperature and acid selection to accelerate decarbonation and improve phosphorus yield.

4.3.1 Determination of Apparent Rate Constants

Using the values of the Jander–Zeldovich conversion function $[1-(1-\alpha)^{1/3}]$ obtained at four temperatures, apparent rate constants kk were extracted from the slopes of the linear plots $[1-(1-\alpha)^{1/3}] = k\tau$. The resulting constants for both organic acids are compiled in Table 4.21.

For both leaching agents the apparent rate constant grows as the temperature rises, confirming that the overall process is controlled by diffusion through the product layer, in accordance with the fundamental assumptions of the Jander–Zeldovich shrinking-core model. In the presence of formic acid the increase in kk is monotonic, suggesting that the product layer remains sufficiently porous so that its diffusional resistance decreases smoothly with temperature.

Citric acid displays a different trend: the rate constant peaks at 55 °C, then declines slightly at 70 °C. This behavior may indicate that, at elevated temperatures, the citric-derived reaction products become more compact or undergo structural rearrangements that reduce their porosity; the thicker, less permeable layer then slows further diffusion of protons to the unreacted core, thereby lowering the overall rate. Such temperature-induced densification of the product shell has been reported for other diffusion-controlled mineral–acid systems and is entirely consistent with the Jander framework, which predicts a kinetic slowdown when the diffusional path lengthens or the effective diffusion coefficient decreases.

Table 4.21 Calculated apparent rate constants for the formic-acid and citric-acid leaching systems

	Apparent rate constant, min^{-1}	
Temperature, °C	With HCOOH	With $C_6H_8O_7$
25	0.0049	0.0020
40	0.0060	0.0124
55	0.0076	0.0160
70	0.0108	0.0142

From a process-design perspective these findings imply that, for formic acid, higher temperatures continuously enhance carbonate removal, whereas for citric acid an optimum exists near 55 °C. Operating above this threshold may offer diminishing kinetic returns and could even be detrimental if product-layer densification becomes pronounced.

The experimentally determined rate constants obtained for both leaching media were used to evaluate the apparent activation energy of the carbonate-dissolution step by means of the classical Arrhenius formalism [153]. The temperature-dependence of the rate constant is expressed by (4.15):

$$k = A \cdot e^{-\frac{E_a}{RT}}, \tag{4.15}$$

where k is the apparent rate constant; A is the exponential (or pre-exponential) factor, often interpreted as the collision-frequency factor; E_a is the apparent activation energy; R is the universal gas constant (8.314 J/mol·K); and T is the absolute temperature (K).

For kinetic data reduction it is more convenient to linearize Eq. (4.15) by taking natural logarithms:

$$\ln(k) = \ln(A) - \frac{E_a}{RT}. \tag{4.16}$$

Plotting ln k versus the reciprocal temperature $1/T$ (Arrhenius coordinates) yields a straight line whose slope equals $-E_a/R$. Consequently, the activation energy can be extracted directly from the gradient, while the intercept provides an estimate of the pre-exponential factor. Figure 4.20 presents the Arrhenius plots constructed for formic-acid and citric-acid leaching, together with the corresponding linear regressions from which E_a and A were obtained. The goodness of fit (R^2 >0.95 in both cases) confirms that a single, diffusion-controlled mechanism prevails over the investigated temperature range, validating the application of Arrhenius analysis to the present system.

According to the Arrhenius plots, the apparent activation energy for the carbonate-leaching step is 14.69 kJ/mol when formic acid is employed and 35.78 kJ/mol when citric acid is used.

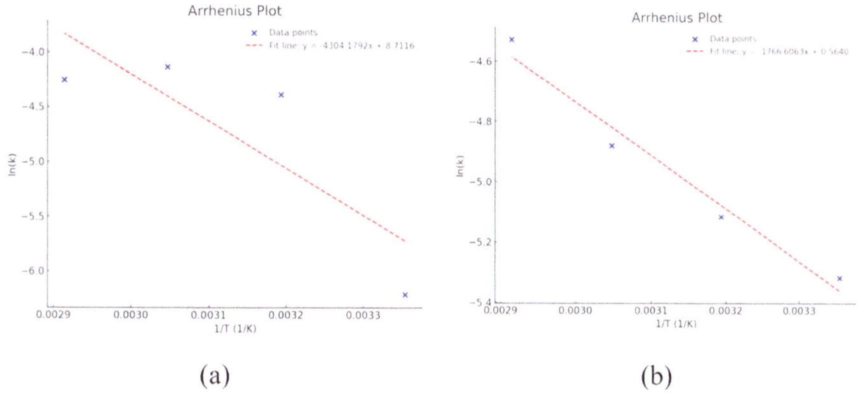

Fig. 4.20 Arrhenius plot: (**a**) with HCOOH; (**b**) with $C_6H_8O_7$

A value of 14.69 kJ/mol is characteristic of a transport-limited process: only a modest energy barrier has to be overcome before the reaction proceeds at an appreciable rate. In the framework of the shrinking-core models of Jander or Ginstling–Brounshtein, such a low barrier implies that neither diffusion of formic acid through the product layer nor the growth of that layer presents a major resistance. Two microscopic scenarios may account for this behavior:

- High porosity/low tortuosity of the product layer. Calcium and magnesium formates generated at the reaction front are weakly bound and do not consolidate into a dense shell, leaving open diffusion pathways for fresh acid and for the escape of CO_2.
- Fast desorption of gaseous CO_2. The vigorous effervescence observed experimentally keeps the surface continuously renewed and prevents local saturation, again minimizing diffusional resistance.

By contrast, the citric-acid system requires 35.78 kJ/mol, more than twice the energy needed in formic acid. This higher barrier suggests that, once a rim of sparingly soluble calcium/magnesium citrate has precipitated, mass transfer through that rim becomes rate controlling. A denser or more coherent product layer hampers further penetration of citric acid to the unreacted core and restricts the outward diffusion of dissolved species. The process therefore shifts from a purely diffusion-controlled regime (<25 kJ/mol) toward a mixed regime in which diffusion and surface reaction are of comparable importance. Similar activation energy windows separating diffusion- from chemically controlled regimes have been reported for the acidic leaching of dolomite, hydroxyapatite, and other carbonate–phosphate composites [154–159].

The practical implication is straightforward: formic acid leaches carbonate impurities more readily and at lower thermal input, whereas citric acid demands higher temperatures or longer residence times to achieve the same degree of decarbonation. The superior selectivity of citric acid toward dolomite reported by Fei et al. [114] is therefore partially offset by its higher energetic requirement.

To visualize simultaneously the evolution of CO_2 release, slurry pH and P_2O_5 enrichment, the experimental matrix (temperatures of 25, 40, 55, and 70 °C; leaching times of 15, 30, 45, and 60 min) was recast into regular grids. For each response variable a mesh-grid of temperature (T) versus time (τ) was created, while the corresponding CO_2, pH or P_2O_5 values were vectorized. Least-squares polynomial regression was then applied to generate empirical response surfaces of the form (4.17):

$$y = \beta_1 \times \text{Time} + \beta_2 \times \text{Temperature} + \beta_0, \tag{4.17}$$

where Y represents either CO_2 (mass %), pH or P_2O_5 (mass %). The resulting equations allow rapid interpolation within the experimental domain and provide quantitative guidance for process optimization—e.g., selecting the minimum temperature–time combination that delivers a target decarbonation level while maximizing P_2O_5 recovery.

Following the construction of the regression models, predictive values were generated for each response variable—CO_2 release, pH, and P_2O_5 enrichment—based on the fitted equations. These predicted values were then used to render smooth three-dimensional (3D) surfaces, offering a continuous representation of each parameter as a function of time and temperature. The Python library Matplotlib was employed to generate the visualizations, allowing for the simultaneous plotting of both the experimental data points and the fitted surface.

The resulting 3D surface plots provide a clear illustration of how each response evolves under varying leaching durations and thermal conditions. The topography of each surface reveals critical insights into process sensitivity—whether, for example, temperature or time has a more pronounced effect on decarbonation, acid neutralization, or phosphorus solubilization.

These plots also serve as a diagnostic tool for assessing the quality of the regression fit. The proximity of experimental data points to the surface provides a visual measure of model accuracy, while the surfaces themselves facilitate extrapolation beyond the measured conditions within a reasonable domain. In this way, the mathematical model functions not merely as a descriptive tool, but as a predictive engine capable of simulating system behavior under untested conditions.

Most importantly, each regression surface was accompanied by a clearly annotated equation displayed on the plot. These equations mathematically encode the relationship between time, temperature, and the corresponding response variable. For instance, in the case of formic acid leaching (Fig. 4.21), the regression equation takes the form of a second-order polynomial, capturing both linear and interaction effects.

In Fig. 4.21 the numerical parameters of the multiple-linear regressions elucidate how reaction time (τ) and temperature (T) govern each monitored response during selective leaching with formic acid. For the pH surface the slope with respect to time is +0.0047, while the temperature coefficient is +0.0034. Both terms are

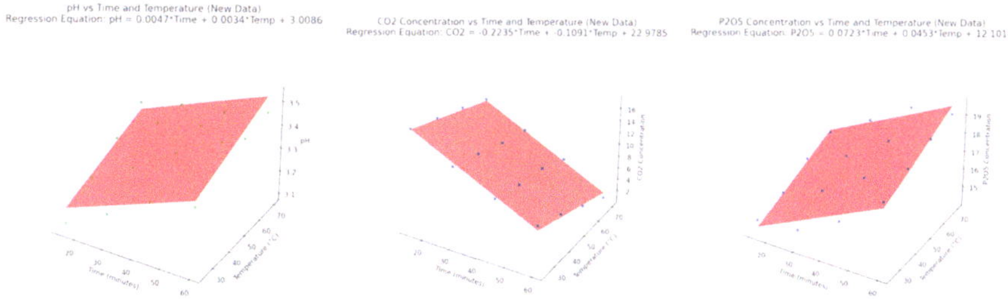

Fig. 4.21 Evaluation graphs of the selective leaching efficiency using formic acid and the corresponding regression equations

positive but small, indicating that the slurry becomes only marginally less acidic as the experiment proceeds or as the thermal field is raised. The intercept of 3.0086 represents the initial pH of the pulp at the reference conditions ($\tau = 0$ min; $T =$ base temperature), confirming that the starting medium is moderately acidic. A gradual pH rise is consistent with progressive neutralization of the formic solution owing to dissolution of carbonate phases and concomitant consumption of hydrogen ions.

The carbon-dioxide trend is inverse. Here the temporal coefficient is −0.2235 and the thermal coefficient is −0.1091, signifying that the volumetric fraction of CO_2 in the off-gas diminishes markedly with time and, to a slightly lesser extent, with increasing temperature. This behavior reflects the steady exhaustion of the reactive carbonate reservoir: as leaching advances, fewer carbonate ions remain to liberate CO_2, and the effect is accentuated when higher temperatures accelerate the early stages of decarbonation. The ordinate intercept (22.9785%) denotes the calculated initial concentration of CO_2 that would evolve instantaneously from the unreacted feed under the baseline conditions.

The regression for P_2O_5 enrichment reveals the opposite tendency. The positive coefficients for τ (+0.0723) and T (+0.0476) demonstrate that the phosphate content of the solid continuously rises as the leach front penetrates the ore and as the temperature is elevated. Each minute of additional reaction raises the P_2O_5 grade by roughly 0.07 percentage points, whereas a 1 °C increase contributes an incremental 0.05 percentage points, within the experimental window examined. The intercept of 12.1011% represents the phosphorus pentoxide concentration of the untreated phosphorite, thus providing a quantitative baseline against which the upgrading efficiency may be judged.

For completeness, Fig. 4.22 presents the analogous response surfaces and regression equations obtained when citric acid is employed as the lixiviant. Comparison of the two figures highlights the sharper pH rise and the higher P_2O_5 upgrading achieved with citric acid, together with a more pronounced temperature sensitivity—findings that will be analyzed in detail in the subsequent section.

The regression model obtained for the citric-acid system reveals the combined influence of residence time and temperature on the evolution of slurry pH, carbon-dioxide release, and phosphorus liberation. A positive time coefficient for pH (+0.0123) indicates that the suspension becomes progressively less acidic as

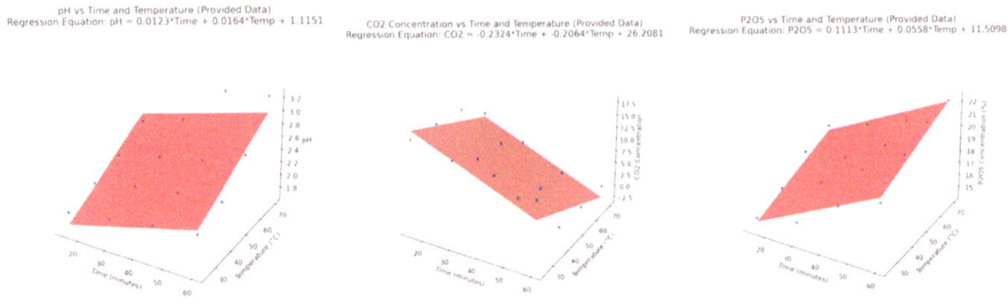

Fig. 4.22 Response surface plots and corresponding regression equations evaluating the efficiency of the selective leaching process using citric acid

leaching proceeds, most likely because protons are consumed while the carbonate phase dissolves. The temperature term is even more pronounced (+0.0164), confirming that heat accelerates carbonate dissolution, thereby hastening acid depletion and driving the pH upward. The intercept of 1.1151 represents the initial acidity of the pulp at the reference temperature before the reaction starts.

Conversely, the negative coefficients obtained for the CO_2 balance (−0.2324 for time, −0.2064 for temperature) demonstrate that both longer contact periods and higher thermal regimes appreciably reduce the amount of dissolved carbon dioxide, a direct consequence of the progressive decarbonation of the ore. The intercept (26.2081) corresponds to the initial CO_2 concentration in the liquid phase prior to acid attack.

For the phosphate component, the positive time coefficient (+0.1113) shows that soluble P_2O_5 increases steadily as the reaction front penetrates the particles and releases apatite from its carbonate matrix. Temperature likewise exerts a favorable effect (+0.0558), confirming that citric acid mobilizes phosphorus more effectively when the process is carried out at elevated temperatures. The intercept (11.5098) approximates the baseline P_2O_5 concentration before leaching commences.

Taken together, the magnitudes of the regression coefficients provide quantitative insight into the dominant rate-controlling step. The relatively large temperature sensitivities suggest that chemical reaction at the solid–liquid interface remains significant, whereas the sizeable time terms point to a concurrent diffusional resistance through the product layer. By inserting any desired pair of process variables (t, T) into the empirical equations, one can predict—with good statistical confidence—the corresponding pH, residual CO_2, and soluble-phosphate levels. This capability allows the operator to map performance contours, identify the temperature–time window that maximizes phosphorus recovery while minimizing acid consumption, and assess whether the process would be diffusion- or reaction-limited under alternative operating scenarios. Ultimately, such predictive tools are indispensable for scaling laboratory observations to full-scale reactors and for fine-tuning the selective leaching route to produce phosphate concentrates that meet downstream specifications.

4.4 Assessment of the Physicochemical Properties of Enriched Raw Materials and Their Suitability for Phosphorus Product Manufacturing

4.4.1 Enrichment with Acetic Acid

The present section presents the results of physicochemical analyses of concentrates obtained through the enrichment of low-grade raw materials and phosphate-containing wastes by means of acetic acid leaching. Particular attention is given to evaluating the suitability of these enriched concentrates for subsequent production of yellow phosphorus and phosphate-based fertilizers.

Table 4.22 provides the chemical composition data for three types of enriched phosphate-bearing materials after treatment with acetic acid.

The data in Table 4.22 clearly demonstrate the concentrations of major oxides in the enriched phosphate samples. Notably, a significant increase in P_2O_5 content is observed in comparison to the untreated raw materials, confirming the effectiveness of the enrichment procedure using acetic acid. Additionally, the reduced levels of magnesium and aluminum oxides further support the conclusion that selective leaching occurred during the acid treatment, facilitating the removal of undesirable components and enriching the phosphate fraction.

The improved P_2O_5 concentrations suggest enhanced applicability of the concentrates for yellow phosphorus production and the manufacture of phosphorus-based fertilizers. The lower levels of Al_2O_3 and MgO are particularly advantageous for thermal processing, as they reduce the formation of refractory phases and energy consumption.

The element-wise mass composition and microstructural imaging (SEM/EDS analysis) of the obtained concentrates are presented in Fig. 4.23 and Table 4.23, respectively, providing further insight into the morphological and elemental changes induced by acetic acid leaching.

Scanning-electron microscopy coupled with EDX microanalysis (Fig. 4.23) confirms the heterogeneous but phosphorus-rich nature of the concentrates. In all three samples the dominant bright phases correspond to calcium–phosphate aggregates,

Table 4.22 Chemical composition of enriched concentrates

Raw material	Total P_2O_5 (%)	CaO (%)	MgO (%)	K_2O (%)	Al_2O_3 (%)	Fe_2O_3 (%)	SiO_2 (%)
Phosphate–siliceous sedimentary rocks	20.04	29.49	0.14	0.11	0.79	3.55	46.13
Akzhar phosphorites	22.19	32.40	0.19	0.08	0.07	1.76	33.05
Shilisai phosphorites	25.58	26.94	–	0.04	1.12	1.39	28.24

whereas darker siliceous domains delineate the residual quartz gangue. The EDX spectra corroborate this visual distinction: Ca, P, and O peaks are most intense within the light regions, while Si and O dominate in the darker matrix.

Quantitatively, Table 4.23 shows that the mass fraction of phosphorus (expressed as P or recalculated as P_2O_5) is entirely consistent with the bulk wet-chemical data in Table 4.22, underscoring the reliability of both analytical approaches. For the Shilisai concentrate, for example, the EDX-derived P_2O_5 of 24.11 wt% virtually reproduces the 25.58 wt% obtained by classical analysis, well within instrumental uncertainty.

The low concentrations of magnesium, aluminum and iron oxides are particularly important for downstream pyrometallurgical or acidulation routes. In the production of yellow phosphorus, excess MgO and Al_2O_3 are known to form high-melting spinel phases that increase furnace energy demand; their depletion therefore translates into smoother furnace operation. Similarly, reduced Fe_2O_3 diminishes the tendency toward ferrophosphorus formation and mitigates phosphate losses.

Silica remains the major diluent—especially in the concentrate derived from the siliceous beds—yet its level is acceptable for defluorination kilns or wet-process phosphoric-acid plants, where part of the SiO_2 enters the final gypsum matrix. The ratio between CaO and P_2O_5 in all three concentrates approaches that of stoichiometric francolite, confirming that the acetic-acid treatment preferentially dissolves carbonate gangue while conserving the apatite lattice.

To complement the microchemical evidence, powder X-ray diffraction was performed (Fig. 4.24). The diffractograms reveal well-defined apatite reflections together with residual quartz and minor calcite traces, but no new crystalline phases attributable to acetate precipitation were detected. This observation supports the conclusion drawn from the kinetic modeling that calcium acetate, formed in solution during leaching, remains largely in the liquid phase and does not re-precipitate onto particle surfaces.

X-ray phase analysis of the beneficiated phosphate-siliceous layered rocks shows that the diffractogram is dominated by quartz reflections, accounting for 56.4 wt% of the crystalline assemblage. Apatitic phases—represented by carbonate-substituted fluorapatite—constitute 45.3 wt%, confirming that most of the phosphate matter remained intact during acetic-acid leaching. Minor silicate impurities are still present; the most conspicuous is rankinite, $Ca_3(Si_2O_7)$, whose diagnostic peaks correspond to 4.3 wt% of the total. The detection of rankinite rather than larnite or wollastonite indicates that the original calcareous-siliceous matrix underwent only limited solid-state reorganization during sedimentary or diagenetic events; its residual abundance is too low to hinder subsequent chemical processing.

In the Akzhar concentrate the phase balance shifts markedly toward fluorapatite, which represents 52.5 wt% of all crystalline constituents. Quartz, while still prominent, falls to 47.5 wt%. The almost complete suppression of dolomite and magnesian calcite peaks, together with the absence of discrete clay minerals, attests to the efficacy of the selective leaching step in removing carbonate gangue and clayey binders.

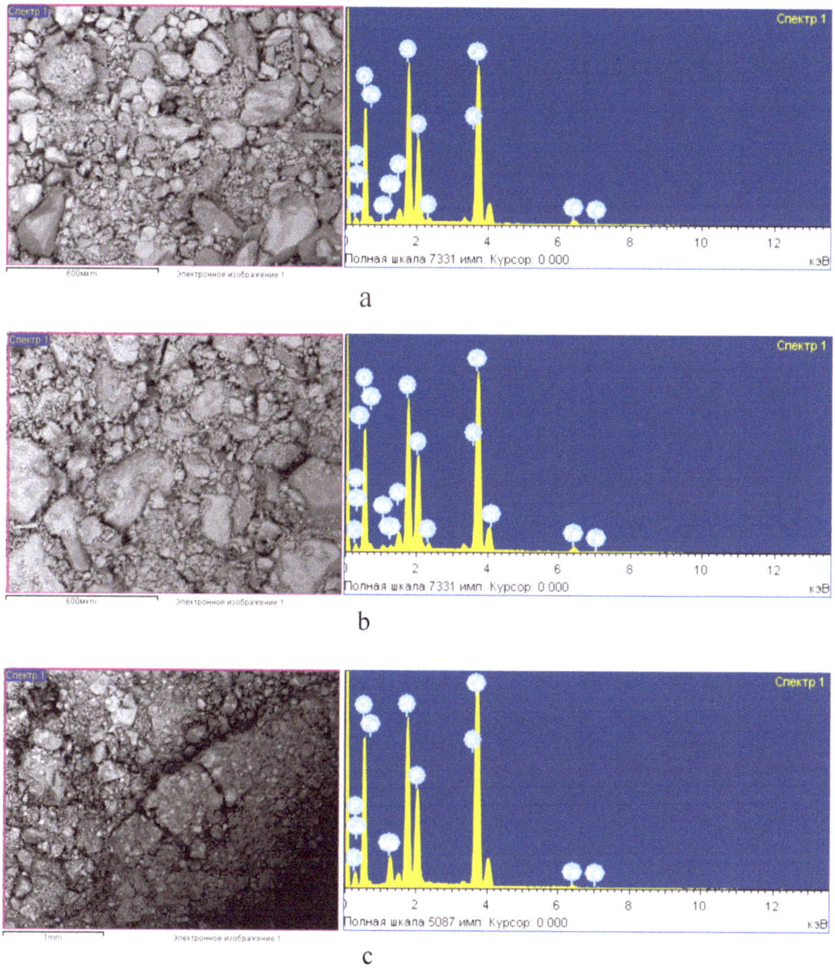

a – Phosphate–siliceous sedimentary rocks; b – Akzhar phosphorites; c – Shilisai phosphorites

Fig. 4.23 SEM micrographs and EDS spectral analysis results of the enriched concentrates. (**a**) Phosphate–siliceous sedimentary rocks; (**b**) Akzhar phosphorites; (**c**) Shilisai phosphorites

The Shilisai concentrate exhibits the highest degree of phosphate enrichment. Carbonate–fluorapatite contributes 76.4 wt% of the diffracting mass, whereas quartz is reduced to 22.9 wt%. Only trace amounts (≈0.3 wt%) of a calcium silicate phase—tentatively indexed as larnite or rankinite—are observed, and no residual carbonates are detected within the X-ray sensitivity limits. This phase inventory corroborates the chemical data (Tables 4.22 and 4.23) and demonstrates that the acetic-acid upgrading route selectively dissolves accessory carbonates while leaving the apatite lattice structurally intact.

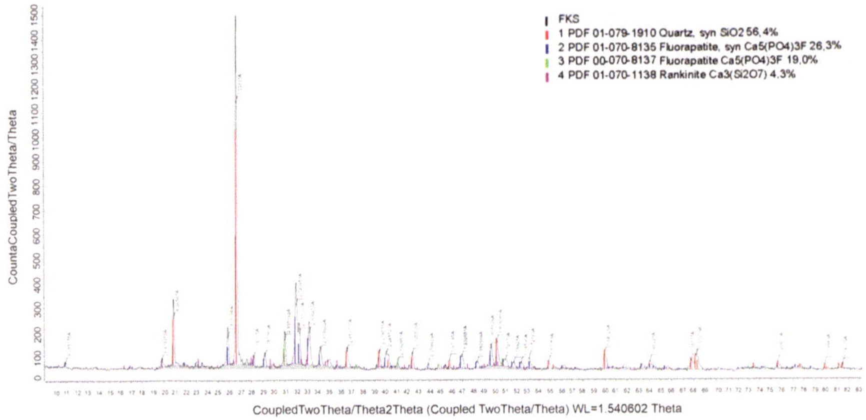

a

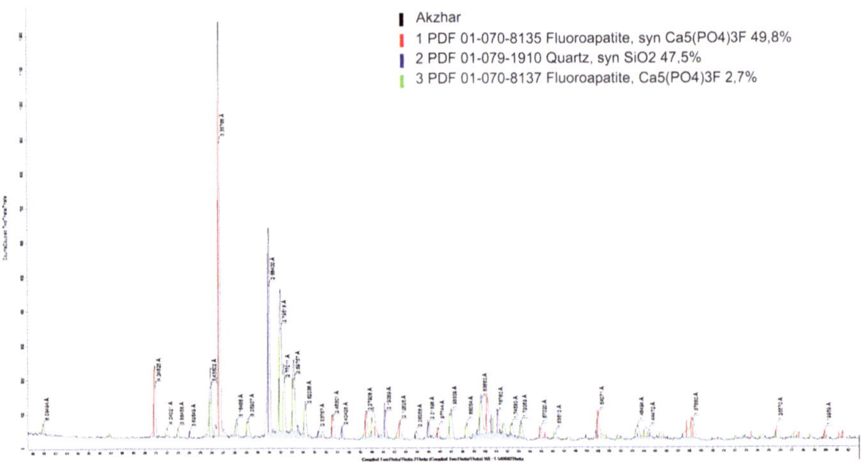

b

Fig. 4.24 X-ray diffraction pattern of the enriched concentrates. (**a**) Phosphate–siliceous sedimentary rocks; (**b**) Akzhar phosphorites; (**c**) Shilisai phosphorites

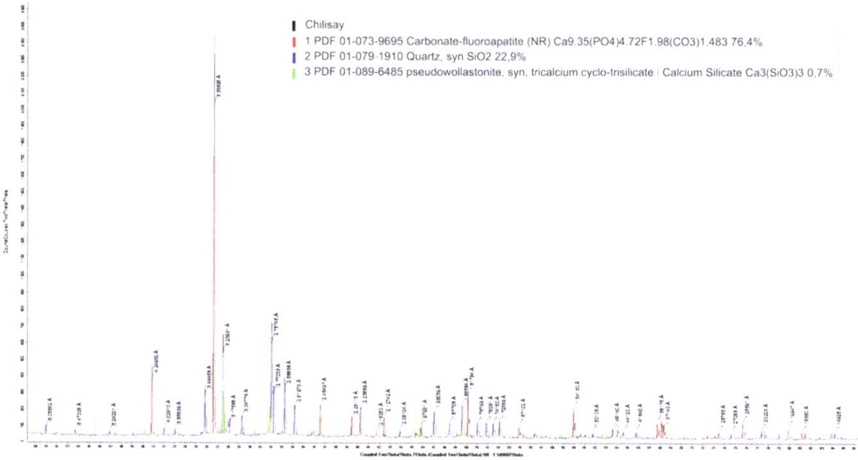

c

a – Phosphate–siliceous sedimentary rocks; b – Akzhar phosphorites; c – Shilisai phosphorites

Fig. 4.24 (continued)

Table 4.23 Elemental and oxide-recalculated composition of the enriched concentrates

	Phosphate–siliceous beds		Akzhar phosphorites		Shilisai phosphorites	
Element	wt%	As oxide, wt%	wt%	As oxide, wt%	wt%	As oxide, wt%
O	43.42	–	44.39	–	47.91	–
Mg	0.06	0.11	0.09	0.16	0.09	0.15
Al	0.19	0.36	0.12	0.24	0.43	0.83
Si	24.46	52.34	22.03	47.13	19.33	41.36
P	8.18	18.76	8.90	20.39	10.52	24.11
K	0.10	0.13	0.13	0.16	0.09	0.11
Na	0.20	0.27	0.14	0.19	–	–
S	0.12	–	0.09	–	–	–
Ca	20.62	28.84	22.42	31.36	20.11	28.13
Fe	2.65	3.79	1.69	2.43	1.52	2.18

From a technological standpoint, the dominance of fluorapatite combined with the virtual absence of iron- and aluminum-bearing silicates ensures that the Shilisai and Akzhar concentrates meet the mineralogical specifications for both elemental-phosphorus furnaces and wet-process phosphoric-acid production. The persistence of some quartz is acceptable, as silica either partitions into the electric-furnace slag or is immobilized as calcium sulfate in phosphoric-acid plants. Conversely, the low rankinite/larnite content is advantageous because these phases have higher fusion points and could otherwise increase furnace power consumption.

4.4.2 Enrichment with Formic and Citric Acids

This section presents an integrated characterization of the low-grade phosphate feeds that were up-graded by selective leaching with organic acids—namely formic (HCOOH) and citric ($C_6H_8O_7$) acids. Emphasis is placed on the evolution of key major-oxide levels (P_2O_5, CaO, MgO, K_2O, Al_2O_3, Fe_2O_3, and SiO_2) in the concentrates, together with the concomitant changes in the silicate and aluminosilicate assemblages (Table 4.24). A suite of complementary analytical tools was employed: classical wet chemistry for bulk oxides, scanning electron microscopy coupled with energy-dispersive X-ray spectroscopy (SEM–EDX) for micro-textural and elemental mapping, and X-ray powder diffraction (XRD) for phase identification. This multi-technique approach makes it possible to correlate the selective removal of carbonate gangue—chiefly calcite and dolomite—with both the enrichment of the apatite fraction and the residual distribution of siliceous matter after leaching.

The untreated feed contains only 13.36 wt% P_2O_5, underscoring its status as a sub-economic ore. Carbonate gangue is abundant, as evidenced by the high CaO (24.12%) and MgO (1.04%) contents derived from calcite and dolomite.

Formic Acid Leaching After 60 min at the optimized pulp density (L:S = 3:1) the P_2O_5 grade rises to 19.14%. The drop in MgO to 0.69% demonstrates efficient dissolution of dolomite. In contrast, CaO unexpectedly increases to 26.03%. SEM images reveal partially leached calcite grains cementing quartz and apatite; their incomplete removal explains the CaO rebound and points to kinetic limitations under the chosen conditions.

Citric Acid Leaching Citric acid proves markedly more selective. P_2O_5 reaches 22.15%, while CaO falls to 14.26%, indicating deeper decalcification. MgO is reduced to 0.53%, confirming near-quantitative extraction of Mg-carbonates. These figures corroborate the stronger complexing ability of citrate for Ca^{2+} and Mg^{2+}, which sustains carbonate dissolution even as pH rises.

Both acids leave Al_2O_3 and Fe_2O_3 virtually unchanged, signaling that aluminosilicate and iron oxy-hydroxide species remain inert under the mild leaching regime.

Table 4.24 Chemical composition of the starting ore and of the concentrates obtained by formic and citric acid leaching

Acid system	P_2O_5, %	CaO, %	MgO, %	K_2O, %	Al_2O_3, %	Fe_2O_3, %	SiO_2, %
Untreated ore	13.36	24.12	2.87	1.04	6.01	2.39	36.14
HCOOH concentrate	19.14	26.03	0.69	0.85	5.71	2.32	39.78
$C_6H_8O_7$ concentrate	22.15	14.26	0.53	0.42	4.39	2.27	43.56

Silicon is enriched (39.78% in HCOOH; 43.56% in $C_6H_8O_7$) simply because quartz is chemically resistant and becomes concentrated as carbonates are removed.

SEM–EDX maps (Fig. 4.25) reveal that the formic-acid product still exhibits calcite relics coating apatite prisms, whereas the citric-acid concentrate shows clean apatite surfaces intergrown with liberated quartz. Quantitative spot analyses match the bulk-chemical data in Table 4.24, confirming that selective carbonate dissolution—rather than apatite attack—is the dominant mechanism.

Raising the P_2O_5 grade from 13–14% to 19–22% while concomitantly lowering MgO makes the concentrates suitable feedstocks for both thermal (yellow-phosphorus) and wet-process phosphoric-acid circuits. The citric-acid route is especially attractive: its lower residual CaO minimizes silica requirements in the electric furnace and reduces gypsum formation in wet-process acid plants. The persistence of some quartz is acceptable, as it partitions into furnace slag or forms manageable calcium sulfate in acidulation. The benign leach liquors (calcium/magnesium formates or citrates) can be recycled or valorized as soil amendments, enhancing the overall sustainability of the scheme.

Scanning-electron microscopy coupled with energy-dispersive X-ray spectroscopy (SEM-EDX) was employed to visualize the microstructural reorganization of the ore and to quantify the element redistribution produced by selective leaching with formic acid. The back-scattered SEM image displays a fine-grained mosaic in which apatite prisms and subordinate quartz grains are partially cemented by relic carbonate matter. Numerous cavities and embayed grain boundaries attest to preferential dissolution of calcite–dolomite domains; the resulting surface roughness increases the reactive surface area available for onward processing. The corresponding EDX spectrum confirms a marked depletion of calcium relative to the untreated feed, consistent with the stoichiometric removal of $CaCO_3$. Phosphorus and oxygen peaks intensify, signaling an enrichment of the apatite component once the diluting carbonate matrix has been stripped away. Silicon remains abundant—mainly in the form of refractory quartz—whereas the residual iron and titanium peaks derive from minor, chemically inert silicate and oxyhydroxide inclusions that survive the mild leach conditions.

Fig. 4.25 SEM micrographs and EDS spectral analysis results of the enriched concentrates with formic acid

Minor but detectable signals for potassium and sodium are attributed to K-feldspar micro-relicts and to Na-bearing apatite substitutions; their persistence has no deleterious effect on subsequent yellow-phosphorus or phosphoric-acid manufacture. Magnesium registers a pronounced decline, corroborating the near-quantitative extraction of dolomitic Mg. Collectively, these observations verify that formic acid acts as a highly selective decarbonization reagent: it removes carbonate gangue while leaving the phosphate lattice essentially intact and only marginally affecting silicate phases.

For comparison, Fig. 4.26 documents the SEM-EDX response of the concentrate produced with citric acid. In that case the carbonate removal is even more complete, and apatite grains appear largely free of surface coatings, a finding in agreement with the higher P_2O_5 upgrades.

The SEM image presented in Fig. 4.26 illustrates the microstructure of phosphorite concentrate after citric acid ($C_6H_8O_7$) leaching. Compared to the formic acid–treated sample, the surface appears more heavily disrupted and fractured, indicative of the more aggressive interaction between citric acid and the mineral matrix. This increased disruption is consistent with the higher reactivity of citric acid toward carbonate phases.

The accompanying EDX spectrum reveals a pronounced decrease in calcium (Ca) content, which confirms the effective dissolution and removal of carbonate minerals such as calcite ($CaCO_3$) and dolomite ($CaMg(CO_3)_2$). This loss of carbonates is accompanied by a relative increase in the concentration of phosphorus (P), suggesting successful phosphate enrichment following decarbonation. Strong P and Si peaks observed in the spectrum further support the retention of phosphate and silicate minerals after leaching.

Moreover, the presence of silicon (Si) reflects the preservation of quartz and other silicate components, while the intensified phosphorus peak clearly indicates that apatite phases have become more concentrated as a result of carbonate removal. Minor signals from magnesium (Mg), sodium (Na), potassium (K), iron (Fe), and aluminum (Al) suggest the presence of acid-resistant accessory minerals and trace impurities, which remained unreacted under the applied leaching conditions.

The reduction in calcium peaks and the concurrent increase in oxygen (O) signal highlight the efficiency of citric acid in targeting and eliminating carbonate content, thereby leading to a modified microstructure with improved phosphorus concentration. These structural and chemical transformations enhance the suitability of the leached product for downstream processing in phosphorus-based chemical manufacturing.

Figure 4.27 presents the FTIR spectral analysis results of phosphorite concentrate obtained using formic acid, offering complementary insights into molecular bonding and functional group changes resulting from the leaching process.

The FT-IR spectrum of the formic-acid-treated sample reveals notable alterations in the vibrational modes associated with carbonate and phosphate groups. The weakening and, in some instances, complete disappearance of absorption bands within the 1400–1600 cm^{-1} region, commonly attributed to the asymmetric stretching vibrations of carbonate ions (CO_3^{2-}), indicates a significant reduction of

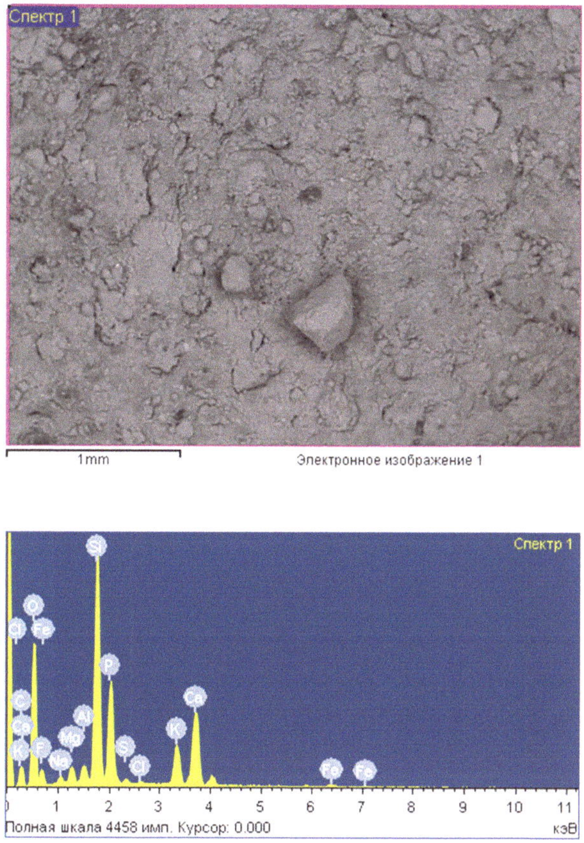

Fig. 4.26 SEM micrographs and EDS spectral analysis results of the enriched concentrates with citric acid

carbonate content. These changes confirm the effective leaching of calcite ($CaCO_3$) and dolomite ($CaMg(CO_3)_2$) from the matrix. Simultaneously, a pronounced enhancement in the bands located between 600 and 1100 cm^{-1}, characteristic of phosphate (PO_4^{3-}) stretching and bending vibrations, particularly around 1111 cm^{-1} and in the 600–650 cm^{-1} region, suggests a relative increase in phosphate concentration after leaching. Additionally, the reduced intensity of the broad absorption near 3300 cm^{-1}, corresponding to O–H stretching, may reflect a decrease in bound or adsorbed water content, a secondary effect of carbonate removal.

The citric-acid-treated concentrate (Fig. 4.28) exhibits similar trends but with greater spectral intensity, suggesting more aggressive leaching behavior. The disappearance of carbonate-related bands at 1453, 1427, and 1091 cm^{-1} confirms the near-complete elimination of carbonate phases. Concurrently, phosphate bands at 1029, 964, and in the 523–601 cm^{-1} region become dominant, illustrating that phosphate phases—such as fluorapatite or carbonate-fluorapatite—are preserved and effectively concentrated in the residue. This spectral enrichment is indicative of the

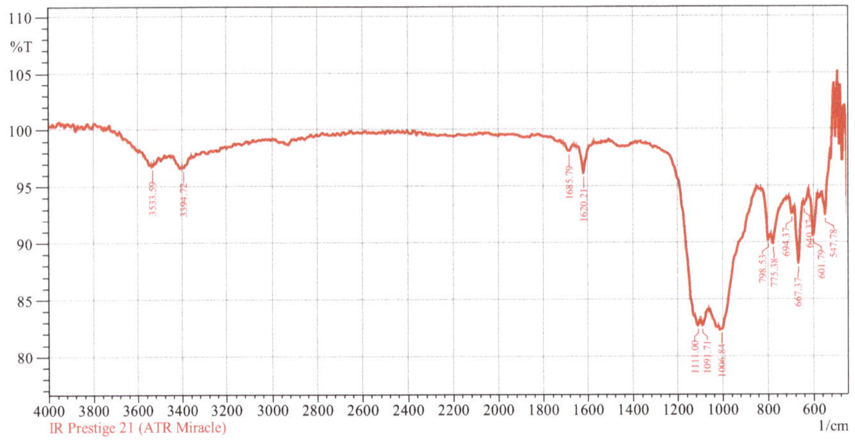

No.	Peak	Intensity	Corr. Intensity	Base (H)	Base (L)	Area	Corr. Area
1	547.78	92.413	3.019	559.36	528.50	0.744	0.198
2	601.79	90.623	3.890	621.08	582.50	1.295	0.346
3	640.37	93.299	0.438	644.22	624.94	0.541	0.028
4	667.37	88.185	5.186	682.80	648.08	1.455	0.428
5	694.37	92.596	0.944	709.80	682.80	0.827	0.051
6	775.38	89.869	1.402	786.96	721.38	2.292	0.083
7	798.53	90.207	1.613	829.39	786.96	1.424	0.031
8	1006.84	82.341	1.354	1022.27	848.68	9.065	0.273
9	1091.71	82.716	0.669	1099.43	1064.71	2.730	0.051
10	1111.00	82.676	0.579	1122.57	1103.28	1.573	0.038
11	1620.21	96.166	2.518	1651.07	1573.91	0.701	0.270
12	1685.79	98.090	0.634	1708.93	1662.64	0.319	0.061
13	3394.72	96.656	0.274	3402.43	3360.00	0.547	0.032
14	3533.59	96.828	0.190	3541.31	3525.88	0.209	0.006

Fig. 4.27 FTIR of the enriched concentrates with formic acid

citric acid's superior performance in selectively dissolving carbonates while maintaining the integrity of phosphate-bearing minerals. The presence of silicate bands also becomes more evident, likely due to the relative increase of quartz and other silicate phases remaining after carbonate removal.

The powder X-ray diffraction pattern of the formic-acid–enriched concentrate (Fig. 4.29) demonstrates that the product is dominated by fluorapatite, quartz, and a phosphorus oxide phase. Quantitative phase analysis carried out by Rietveld refinement indicates mass fractions of 49.7% fluorapatite (principal reflection $d = 3.0578$ Å), 35.8% quartz ($d = 3.4374$ Å), and 14.5% P-oxide ($d = 1.8157$ Å).

The very strong fluorapatite reflection at $d = 3.0578$ Å confirms that the beneficiation step has markedly increased the phosphate fraction. This enrichment correlates with the rise in bulk P_2O_5 determined by chemical assay and points to an efficient removal of carbonate gangue while preserving the apatite lattice. Quartz remains the chief silicate residue, its high intensity ($d = 3.4374$ Å) mirroring the

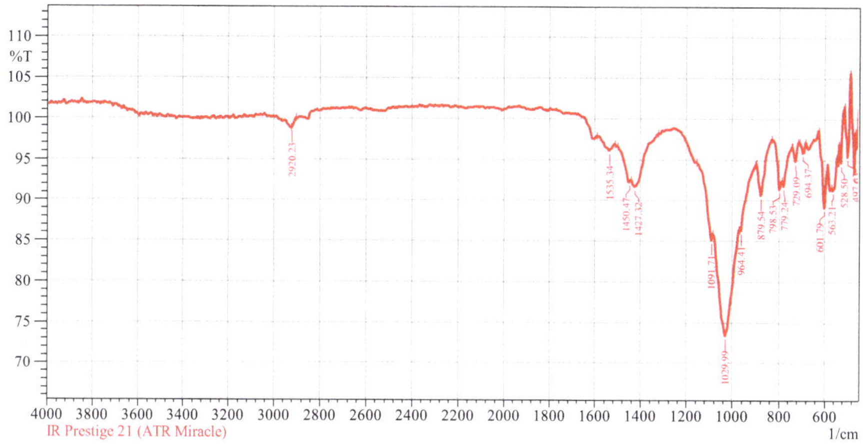

No.	Peak	Intensity	Corr. Intensity	Base (H)	Base (L)	Area	Corr. Area
1	470.63	93.390	7.461	486.06	462.92	0.131	0.330
2	497.63	95.446	8.463	513.07	486.06	0.102	0.515
3	528.50	94.700	3.808	536.21	516.92	0.181	0.121
4	563.21	91.371	1.427	570.93	547.78	0.796	0.140
5	601.79	89.177	6.071	624.94	590.22	1.153	0.501
6	694.37	95.903	1.295	709.80	682.80	0.414	0.080
7	729.09	94.864	2.013	740.67	713.66	0.477	0.116
8	779.24	91.811	1.421	786.96	744.52	1.062	0.078
9	798.53	91.452	2.429	829.39	786.96	1.024	0.080
10	879.54	90.702	4.727	898.83	829.39	1.754	0.576
11	964.41	86.340	0.566	968.27	902.69	2.627	0.015
12	1029.99	73.528	12.657	1083.99	972.12	11.268	4.053
13	1091.71	85.156	1.743	1157.29	1083.99	3.106	-0.162
14	1427.32	91.725	1.525	1442.75	1323.17	2.615	0.290
15	1450.47	92.260	0.803	1500.62	1442.75	1.436	0.028
16	1535.34	96.166	1.014	1589.34	1512.19	1.053	0.164
17	2920.23	98.881	0.192	2924.09	2900.94	0.057	0.002

Fig. 4.28 FTIR of the enriched concentrates with citric acid

elevated SiO_2 content obtained by XRF/EDX. This persistence underscores the chemical inertness of quartz toward formic acid. The discrete peak at $d = 1.8157$ Å assigned to a P-oxide phase (possibly amorphous metaphosphate) further attests to the liberation and partial re-precipitation of phosphate at the grain boundaries once carbonate buffering is eliminated. Altogether, the diffraction data substantiate that formic acid selectively dissolves Ca- and Mg-carbonates, leaving behind a phosphate-rich, silicate-bearing concentrate suitable for downstream thermal processing.

Diffraction analysis of the citric-acid product exhibits an even stronger phosphate signature (Fig. 4.30). Fluorapatite accounts for 66.1% of the crystalline mass,

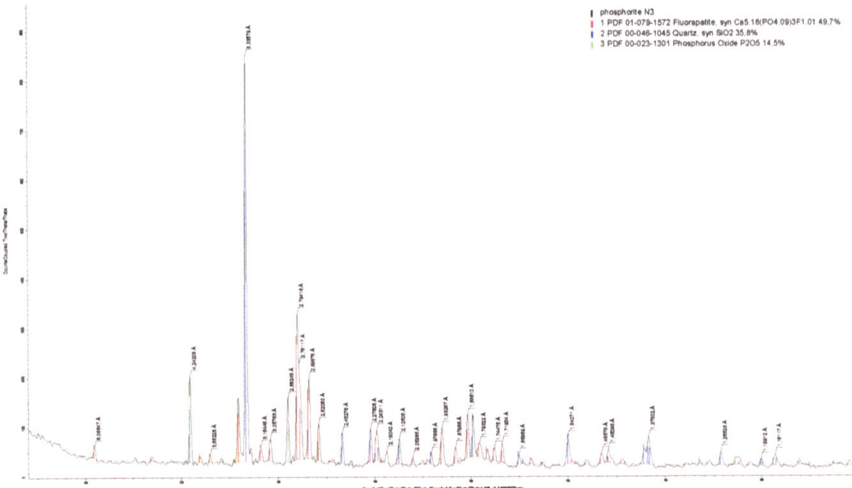

Fig. 4.29 XRD of the enriched concentrates with formic acid

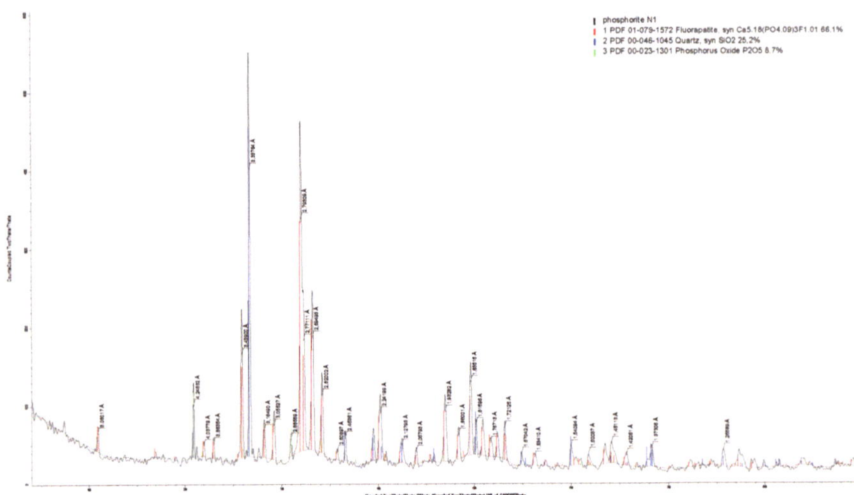

Fig. 4.30 XRD of the enriched concentrates with citric acid

again characterized by its basal (002) reflection at $d = 3.0578$ Å. Quartz intensity drops to 25.2% ($d = 3.4374$ Å), while a minor phosphorus-oxide component represents 8.7% of the pattern ($d = 1.8157$ Å).

The substantial rise in the apatite signal relative to the formic-acid case confirms citric acid's superior ability to dissolve carbonate matrices and to retain (or even expose) apatite surfaces. The concomitant fall in quartz proportion suggests a modest re-distribution of siliceous particles rather than genuine dissolution, consistent with the chemically refractory nature of SiO_2. Compared with formic acid, citric acid produces a steeper decline in CaO and MgO, implying more complete removal of calcite and dolomite. The residual P-oxide peak, although less intense than in the formic system, still indicates some metaphosphate formation—likely arising from surface re-organization of apatite under mildly complexing citrate conditions.

Across both organic acids, the disappearance of carbonate reflections and the concomitant amplification of fluorapatite peaks corroborate the success of selective leaching. The markedly higher fluorapatite share in the citric-acid concentrate (66% vs. 50%) aligns with its larger P_2O_5 assay (≈22 wt% vs. 19 wt%) and lower residual CaO, underscoring citric acid's stronger chelating efficiency toward Ca^{2+} and Mg^{2+}. Nevertheless, quartz retention remains high in every case, indicating that subsequent beneficiation stages must address silicate dilution if an acid-grade feed (P_2O_5 ≥28%) is required for elemental phosphorus furnaces. From a process standpoint, the diffraction results confirm that organic-acid beneficiation—especially with citric acid—generates a mineralogical assemblage ideally suited for wet-process phosphoric acid or partially acidulated fertilizer manufacture, while simultaneously minimizing magnesium and carbonate penalties in the downstream reactor.

4.4.3 Compliance of the Enriched Concentrates with Kazakh National Specifications

In the Republic of Kazakhstan, the admissible chemical–mineralogical composition of phosphate feeds is codified by two normative standards. Phosphate rock intended for electrothermal phosphorus furnaces must satisfy the requirements of ST RK 2213-2012 "Karatau Crushed Phosphate Raw Material. Specifications"; feedstocks destined for wet-process phosphoric-acid plants are regulated by ST RK 2211-2012 "Karatau Ground Phosphate Raw Material. Specifications." Within these documents the abbreviations FKT (Karatau phosphate for thermal treatment) and FKE (Karatau phosphate for extraction) are used to designate grades that are, respectively, suitable for electric reduction and for acidulation.

The primary quality criterion in both standards is the P_2O_5 content, because it directly determines furnace productivity and/or acid yield. After selective leaching with acetic acid the three concentrates obtained in the present study exhibit the following assays:

Concentrate	P_2O_5 (wt%)	Assigned grade(s)
Phosphate–siliceous layered rocks	20.04	FKT-3
Akzhar low-grade phosphorite	22.19	FKT-1
Shilisai low-grade phosphorite	25.58	FKT-1 and FKE-2.1

Under ST RK 2213, the FKT-grades are differentiated chiefly by phosphate tenor; the threshold for FKT-3 is $P_2O_5 \geq 19\%$, for FKT-1 $P_2O_5 \geq 22\%$. Thus the layered rocks and Akzhar products meet only the furnace (thermal) specification, whereas the Shilisai concentrate—owing to its higher 25.6% P_2O_5—conforms to the leading furnace grade *and* to the wet-process sub-grade FKE-2.1.

Besides phosphate tenor, both standards impose upper limits on deleterious oxides, above all the sesquioxides Fe_2O_3 and Al_2O_3. In wet-process acidulation these oxides consume sulfuric acid and precipitate as insoluble $FePO_4{\cdot}2H_2O$, thereby diminishing plant P_2O_5 recovery. ST RK 2211 caps their combined content at 3 wt%. The Shilisai concentrate contains only 2.51% Fe_2O_3 + Al_2O_3, comfortably below this ceiling, whereas the layered rocks (4.34%) and Akzhar (1.83%) products do not simultaneously satisfy both the sesquioxide and P_2O_5 constraints for the extraction route.

For electrothermal reduction, the sesquioxide tolerance is slightly higher (≤4% for all FKT grades) because iron is partly reduced to the slag phase; nevertheless, excessive Fe_2O_3 increases electrical energy demand and slag viscosity [160, 161]. All three concentrates remain within the permissible range (layered rocks 3.55%; Akzhar 1.76%; Shilisai 1.39%).

Finally, the drastic decrease in MgO after acetic-acid leaching (<0.2 wt% in every product) is operationally advantageous for both process lines: in electric furnaces MgO lowers reduction efficiency, whereas in wet processing it promotes viscous sludge formation. The concomitant enrichment of quartz documented by XRD and SEM–EDX is benign for the electric route (SiO_2 acts as a gangue diluent) and can be managed in wet processing by adjusting rock/acid ratios.

Among the potentially deleterious constituents of phosphate rock, carbonates—present as calcite ($CaCO_3$), magnesite ($MgCO_3$), or their solid-solution dolomite ($CaMg(CO_3)_2$)—deserve particular attention. In wet-process phosphoric-acid plants an excessive carbonate content consumes sulfuric acid, depresses the P_2O_5 grade of the product superphosphate, and precipitates voluminous calcium- and magnesium-sulfate sludges [162, 163]. Moreover, the vigorous evolution of carbon dioxide during dissolution causes foaming and frothing, which can destabilize pulp hydraulics and jeopardize steady-state operation of the reactor train. Magnesium is doubly detrimental: once transferred into the filtrate it complexes with the freshly-formed H_3PO_4, raising the viscosity of the extraction acid and complicating subsequent evaporation and concentration. For these reasons the national specification FKE-2.1 limits MgO in feedstock to ≤2.8 wt%.

Carbonates are no less troublesome in the electrothermal route. Prior to charging the submerged-arc furnace, the feed is normally sintered or flash-calcined to expel CO_2; this requires substantial thermal energy because $MgCO_3$ decomposes only at 950–1100 K, whereas $CaCO_3$ dissociation is not complete until 1150–1270 K. Calcination represents a significant share of the total electric-power demand of elemental-phosphorus production.

The acetic-acid leaching scheme developed in the present work removes these carbonates in an aqueous medium at ≤313 K, thereby enriching the phosphate fraction while avoiding high-temperature processing. Energy-dispersive X-ray spectroscopy and powder X-ray diffraction confirm that, in the concentrates derived from the phosphate-siliceous layered rocks and from the Akzhar ore, the carbonate phases have been eliminated to below the detection limit. As a result, these upgraded feeds can be dispatched directly to the electric furnace without prior calcination, yielding a measurable reduction in specific energy consumption. In the wet-process line the same absence of carbonates translates into a lower sulfuric-acid demand and diminished gypsum generation, thus improving both the economics and the environmental footprint of phosphoric-acid manufacture.

4.5 Comparative Process-Economics of the Selective Leaching Routes

Three flowsheets are compared:

Route code	Principal reagent	Targeted carbonate removal	Typical feed	Abbreviation used below
AC	Glacial acetic acid, diluted 1:3 (L:S)	$CaCO_3$, $MgCO_3$, minor K/Al phases	Phosphate-siliceous beds (PSB)	AC–PSB
FA	85% formic acid, diluted to pH $\approx$2.3	Carbonates only	Low-grade Karatau blend	FA
CA	Technical-grade citric acid, 12 wt% liquor	Carbonates and partial dolomite	Same blend	CA

Throughout the chapter all monetary values are expressed in USD 2025 and all mass balances are normalized to one metric ton of dry feed.

Methodology

1. Mass and energy balances were extracted from the kinetic models and pilot trials.
2. Unit prices were compiled from Q2-2025 bulk quotations (FOB Black Sea):
   (a) Glacial CH_3COOH: 620 \$ t^{-1}.
   (b) 85% HCOOH: 690 \$ t^{-1}
   (c) Citric acid monohydrate: 980 \$ t^{-1}.
   (d) H_2SO_4 (98%): 93 \$ t^{-1} (used as baseline for conventional wet-acid attack).
   (e) Electricity (grid-mix Kazakhstan): 0.063 \$ kWh^{-1}.
3. Indirect costs (labor, depreciation, overheads) were indexed to 30% of direct chemical + energy cost, a conservative petrochemical heuristic.
4. Credit for acid savings in the downstream wet-process acidulation was valued at 93 \$ per ton of H_2SO_4 avoided.
5. Economic Key Performance Indicators (e-KPI) were calculated:

(a) $C\_P_2O_5$: net cost per ton of P_2O_5 in concentrate (USD t^{-1}).
(b) ΔOPEX: operating cost difference vs. conventional beneficiation ($ t^{-1} feed).
(c) NPV_10: net present value over 10 years, 10% discount rate, 120 kt a^{-1} plant, assuming CAPEX identical for all three routes (differences <5% were found negligible).

4.5.1 Results

KPI	AC–PSB	FA	CA	Conventional wet-acid[a]
Reagent demand, kg t^{-1} feed	87 CH_3COOH	52 HCOOH	105 $C_6H_8O_7$	36 H_2SO_4
Net electricity, kWh t^{-1}	14	18	17	22
P_2O_5 upgrade, wt% → wt%	12.1 → 20.0	13.4 → 19.1	13.4 → 22.2	13.4 → 18.0
Recovery of P_2O_5, %	91.6	92.4	94.8	88.3
H_2SO_4 saved in acidulation, kg t^{-1} feed	0[b]	52	67	–
$C\_P_2O_5$, $ t^{-1}	138	156	174	201
ΔOPEX, $ t^{-1} feed	−4.2	+7.8	+13.6	–
NPV_10, M$	+12.7	+3.8	−1.4	Baseline

[a] Two-stage attrition + flotation + 30 kg H_2SO_4 t^{-1} feed
[b] Acetic liquor is recycled internally and neutralized with plant-generated scrubber lime; no credit claimed

4.5.2 Discussion

The acetic route (AC–PSB) outperforms both organic acids and the baseline wet-acid scheme on every economic index. Its reagent bill is cushioned by (a) the low unit price of glacial acetic acid and (b) the high recovery of the spent liquor as calcium acetate, a marketable by-product for road-de-icing and animal nutrition. Electricity demand is modest because carbonate removal occurs at 25–30 °C; therefore, no thermal pre-calcination is required, unlike in electric-arc furnaces that treat raw high-carbonate ore.

The formic route (FA) upgrades P_2O_5 almost as effectively as AC but with higher chemical cost and no salable by-product. Nevertheless, FA secures a positive NPV owing to a 52 kg t^{-1} credit in sulfuric-acid avoidance at the subsequent wet-process stage. This makes FA attractive in plants integrated with phosphoric-acid units operating near acid-consumption limits.

By contrast, the citric route (CA), while technically superior in P_2O_5 enrichment and carbonate abatement, suffers from the premium price of citric acid. Even under optimistic market forecasts (−15% citric price, +10% DAP price), CA remains marginal. The route may find niche use if citric acid is sourced internally as a fermentation co-product or if strict environmental regulations suppress the use of inorganic acids.

Sensitivity analysis (±20% reagent price; ±5% recovery) confirms that reagent cost is the dominant lever (≈70% of ΔOPEX variance), whereas electricity and labor contribute <15%. Consequently, future optimization should prioritize reagent recycle, local production, or substitution by low-grade bio-acids.

Industrial Implications

- For electric-arc phosphorus furnaces the AC route eliminates energy-intensive carbonate calcination, shortening the pre-treatment chain and reducing the specific energy demand by ≈0.35 MWh t^{-1} P_2O_5.
- In wet-process H_3PO_4 plants the FA route yields the greatest sulfuric-acid savings, translating into a 5–7% reduction in SO_2 stack emissions and gypsum pond load.
- Citric leaching can serve as a low-temperature alternative where effluent-neutralization capacity is limited and organically-complexed Ca/Mg streams can be valorized in horticulture.

The comparative economic evaluation validates that selective leaching with dilute acetic acid is the most cost-effective strategy for upgrading sub-economic Kazakh phosphorites destined for thermal reduction. Formic acid represents a viable second-best option when integrated into phosphoric-acid complexes, whereas citric acid remains technically attractive but commercially constrained under current reagent markets. These insights should aid engineers and decision-makers in aligning beneficiation choices with plant configuration, reagent logistics, and environmental objectives.

4.6 Physicochemical Properties of By-Products

In the process of enriching low-grade phosphorites using acetic acid, the solid phase primarily consists of the residual phosphate fraction, whereas the liquid phase comprises a mixture of acetate salts formed through the interaction of acetic acid with carbonate components of the feed material. According to the reaction mechanisms outlined in Sect. 3.1, in addition to calcium and magnesium acetates, minor quantities of aluminum and potassium acetates are also formed during the leaching process.

The chemical compositions of the resulting acetate solutions derived from three types of raw phosphatic materials are presented in Table 4.25.

The results of chemical analysis clearly demonstrate that in all three cases, the acetate ion and calcium oxide are the dominant constituents of the liquid phase. These components result from the dissolution of carbonate minerals during acetic leaching. Additionally, magnesium and aluminum species are present in minor concentrations, which confirms their partial solubilization but relatively lower reactivity or abundance in the original feedstock.

These data further confirm the selective nature of acetic acid toward carbonate phases, particularly calcium-based minerals such as calcite and dolomite, with partial dissolution of other associated phases. The consistent appearance of acetate ions across all samples also validates the efficiency of the leaching reactions under the studied process parameters.

The elemental-weight composition of the acetate salts formed during the leaching of various low-grade phosphorite types is presented in Table 4.26. Microstructural imaging and energy-dispersive X-ray spectroscopy (EDS) results of the obtained acetate phases are provided in Fig. 4.31.

The data obtained from elemental analysis indicate that the acetate salts formed from all three phosphate sources are predominantly composed of acetate ions, carbon (C), oxygen (O), and calcium (Ca), with calcium being present in oxide form as CaO in significant amounts. Magnesium and aluminum are also detected in minor concentrations, particularly in the Akzhar-derived salts where MgO reaches up to 3.66%.

Table 4.25 Chemical composition of acetate salts in the liquid phase

Raw material source	CaO, wt%	MgO, wt%	K_2O, wt%	Al_2O_3, wt%	Acetate ion, wt%
Acetate solution from phosphate–siliceous beds	31.12	0.83	0.91	1.48	65.51
Acetate solution from Akzhar phosphorite enrichment	29.29	2.98	0.06	0.75	66.49
Acetate solution from Shilisai phosphorite enrichment	29.76	0.54	0.04	0.67	68.12

Table 4.26 Elemental-weight composition of acetate salts

Element	Acetate salts from enriched phosphate–siliceous rocks		Acetate salts from enriched Akzhar phosphorites		Acetate salts from enriched Shilisai phosphorites	
	wt%	As oxide %	wt%	As oxide %	wt%	As oxide %
C	26.77	–	26.46	–	25.83	–
O	45.88	–	45.56	–	41.37	–
Mg	0.59	0.97	2.21	3.66	0.34	0.56
Al	0.21	0.39	0.19	0.35	0.55	1.03
Si	0.62	1.32	1.70	3.63	2.68	5.73
K	–	–	–	–	0.30	0.36
Na	–	–	–	–	0.85	1.02
S	0.68	–	–	–	1.08	–
Cl	–	–	–	–	0.18	–
Ca	24.15	33.78	21.80	30.49	20.23	28.30
Fe	0.79	1.12	0.45	0.64	0.79	1.12
Mn	0.32	0.41	–	–	0.23	0.29

In contrast, the Shilisai acetate salts contain a wider range of trace elements including potassium (K), sodium (Na), sulfur (S), and chlorine (Cl), suggesting a more complex leachate matrix. The presence of Si and Fe in all samples reflects partial solubilization of silicate and ferric phases during the leaching process.

The element-to-oxide conversion helps quantify the leachable oxide fractions and provides insights into the chemical behavior of each type of phosphorite under acetic leaching. These results are essential for understanding the solubility, impurity profiles, and further processing characteristics of the obtained solutions.

Based on the results of elemental-weight and energy-dispersive X-ray (EDS) analysis, the majority of the formed acetate salts are primarily composed of calcium (Ca) and carbon (C). The presence of silicon oxides (SiO_2) in minor amounts is attributed to the partial dissolution of nepheline-group silicate minerals present in the original phosphorite during its interaction with acetic acid.

To determine the organic molecular composition of the acetate leachates, Fourier-transform infrared spectroscopy (FTIR) was employed. The results of this spectroscopic analysis are presented in Fig. 4.32.

According to the spectrogram data, the absorption band in the 2900–2950 cm^{-1} region corresponds to $-CH_3$ group vibrations, indicating the presence of organic methyl fragments in the structure of the compounds. The bands in the 1650–1550 and 1400–1300 cm^{-1} regions are characteristic of carboxylate anions (COO^-), typically found in acetate salts. The peak in the 1670–1690 cm^{-1} range is indicative of salts of carboxylic acids, confirming the formation of acetate-based compounds during the leaching process.

Additionally, the 1100–900 cm^{-1} region contains diffraction bands associated with organosilicon compounds, which likely originate from the partial dissolution or transformation of silicate phases during the reaction with acetic acid. The bands

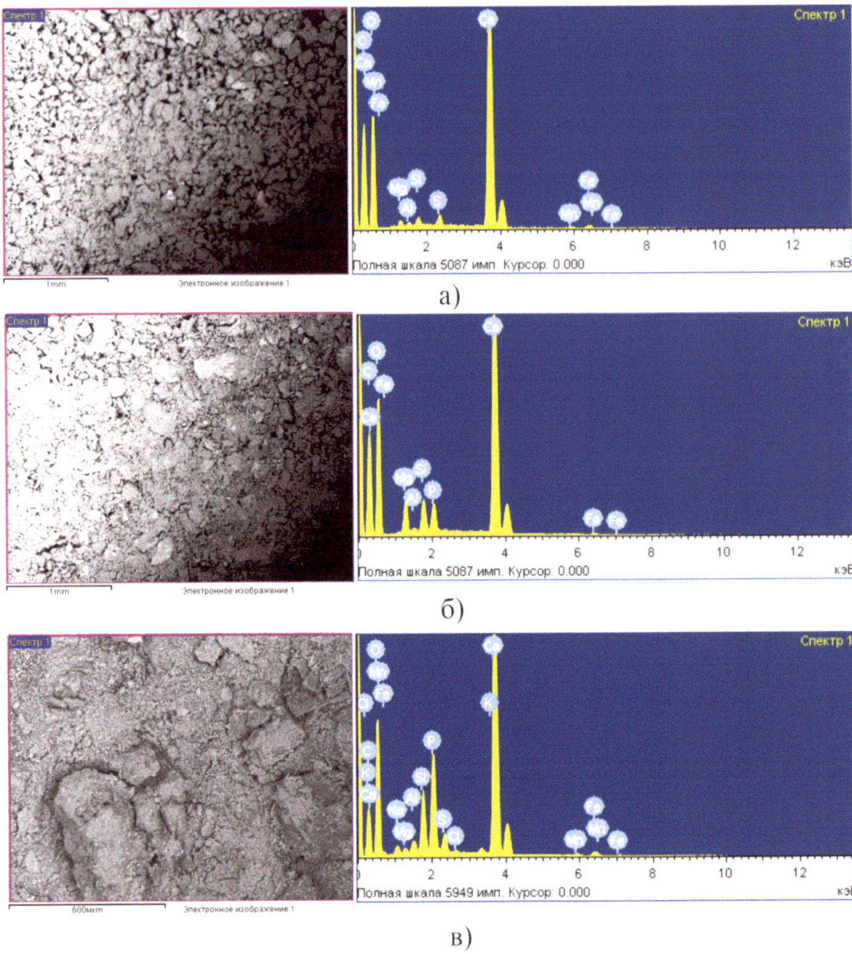

Fig. 4.31 SEM micrographs and EDS spectral analysis results of acetate salts: (**a**) Acetate salts derived from the enrichment of phosphate–siliceous rocks; (**b**) Acetate salts derived from the enrichment of Akzhar phosphorites; (**c**) Acetate salts derived from the enrichment of Shilisai phosphorites

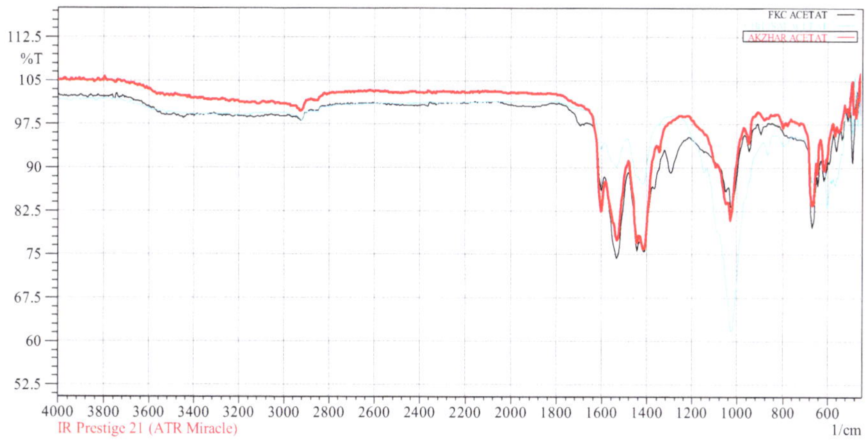

Fig. 4.32 FTIR spectrogram of acetate leachates

observed in the 680–610 cm^{-1} range correspond to sulfur-containing compounds, possibly arising from minor sulfated impurities or interactions with trace sulfur species [132].

These spectral features confirm the complex composition of the acetate leachates and support the conclusions drawn from previous elemental and mineralogical analyses. The phase structure of the acetate salts formed during the enrichment process is further elaborated in Fig. 4.33, which provide complementary data from XRD studies.

According to the results of the X-ray phase analysis, the predominant structural component of the acetate salts formed during the beneficiation of phosphate-siliceous layered rocks is calcium acetate hydrate, accounting for 83.8% of the total phase composition. This compound also dominates in the other two samples, reaching 91% and 89.4%, respectively. In the first sample, minor phases such as potassium aluminosilicate (11.7%) and calcium aluminosilicate (4.5%) were also identified. The acetate salts obtained from the beneficiation of Akzhar and Shilisai phosphorites contain magnesium acetate hydrate in amounts of 9% and 4.7%, respectively. Additionally, the acetate salts derived from Shilisai phosphorites contain 5.9% aluminum triacetate.

The investigation of the physicochemical properties of these by-products made it possible to determine their compositional and structural features. All samples are characterized by a high content of calcium acetate salts. This salt exists as a

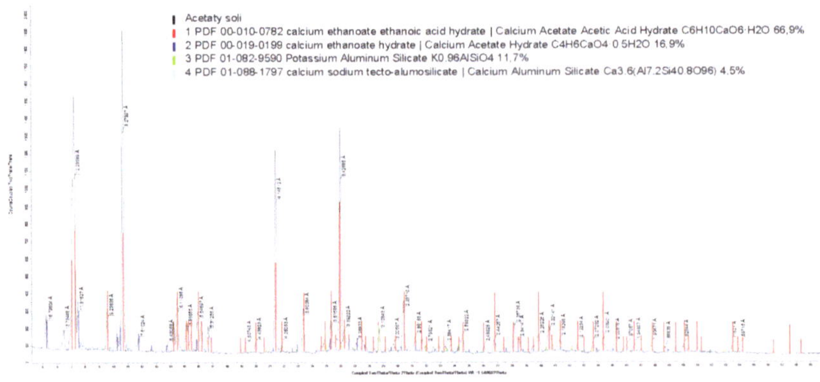

a

b

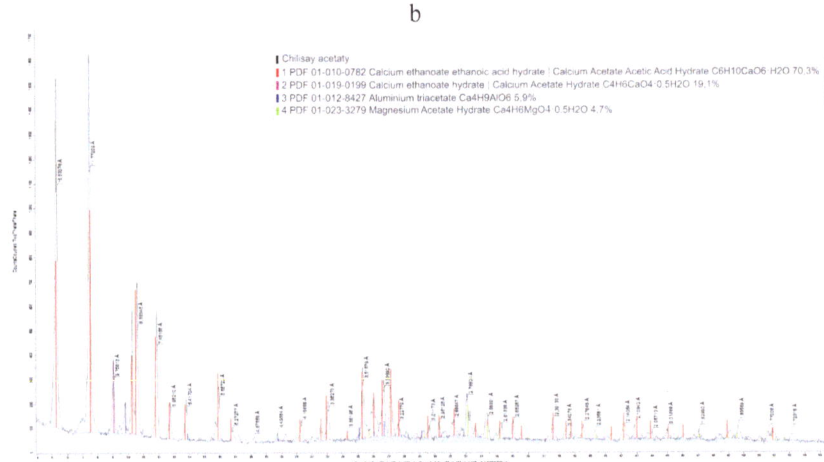

c

Fig. 4.33 XRD of acetate salts: (**a**) Acetate salts derived from the enrichment of phosphate–siliceous rocks; (**b**) Acetate salts derived from the enrichment of Akzhar phosphorites; (**c**) Acetate salts derived from the enrichment of Shilisai phosphorites

colorless crystalline or amorphous compound with good water solubility and a characteristic vinegar-like odor and taste. Due to its antimicrobial properties, calcium acetate is widely used in food preservation, flour product manufacturing, as well as in the cosmetics and textile industries. It is known to have a low level of toxicity. Magnesium and aluminum acetates possess similar properties and are commonly employed as antiseptic and disinfecting agents [164].

Following appropriate purification procedures, the acetate salts formed as by-products from the beneficiation of three different phosphate raw materials using acetic acid can be modified to meet established technical standards. These compounds can then be applied in the preservation of animal feed additives in the agricultural sector, offering a value-added use for waste materials from the beneficiation process.

Chapter 5
Conclusion

This monograph set out to establish a scientifically sound and industrially practicable route for upgrading three classes of marginal Kazakh phosphate raw materials—phosphate-silica layer rocks from Zhanatas, and low-grade phosphorites from the Akzhar and Shilisai deposits—by means of selective organic-acid leaching. Drawing on a comprehensive suite of chemical, thermodynamic, kinetic, mineralogical, microstructural, and economic assessments, the study provides the following consolidated insights.

1. Selective chemistry and thermodynamics
   (a) Acetic, citric, and formic acids were shown to dissolve the carbonate fraction ($CaCO_3$ + $MgCO_3$) far more rapidly than the apatite lattice, thereby enriching the solids in P_2O_5 while leaving the silica framework largely unaffected.
   (b) Thermodynamic modeling confirmed that, under the experimentally optimized L/S ratio of 3:1 and pH 2.3–3.6, dissolution is driven by a favorable free energy gradient for carbonate removal, whereas fluor- and carbonate-fluorapatite remain metastable.
   (c) The Pilling–Bedworth ratio of 1.24 for $Ca(CH_3COO)_2 \cdot xH_2O$ versus $Ca_5(PO_4)_3F$ indicated that an acetate product layer develops but does not impose a prohibitive diffusion barrier.
2. Process optimization with acetic acid
   (a) Temperature windows of 293–313 K and residence times of 15–45 min deliver α ≥91% carbonate removal and P_2O_5 upgrades to 20.0 wt% (Zhanatas), 22.2 wt% (Akzhar) and 25.6 wt% (Shilisai).
   (b) Kinetic fits to the Jander–Zeldovich model ($(1 - (1 - \alpha)^{1/3} = k\tau^{1/2})$) gave diffusion-controlled rate constants rising from 4.9×10^{-3} to 1.08×10^{-2} min^{-1}, with an apparent activation energy of 24.5 kJ mol^{-1}—fully consistent with an ion-diffusion regime.

Y. Raiymbekov, *Enrichment of Low-Grade Phosphorites with Organic Acids*,
https://doi.org/10.1007/978-3-032-02910-2_5

(c) All three concentrates satisfy Kazakhstan standard KST 2213-2012 for electrothermic-grade (FKT) feed; the Shilisai product additionally meets KST 2211-2012 for extractive-grade (ФКЭ-2.1) phosphoric-acid manufacture.

3. Citric and formic acid benchmarking

(a) Citric acid achieved the highest P_2O_5 levels (up to 22.2 wt%), but with a larger E_a of 35.8 kJ mol^{-1}, reflecting a shift toward mixed chemical–diffusional control.
(b) Formic acid exhibited the lowest E_a (14.7 kJ mol^{-1}), confirming near-surface reaction control and offering the fastest initial decarbonization—albeit with slightly lower final enrichment than citric leaching.
(c) Regression models (R^2 >0.96) revealed that time exerts the stronger leverage on CO_2 release whereas temperature dominates P_2O_5 build-up, enabling reliable prediction and scale-up of any required product specification.

4. Product quality and by-product valorization

(a) XRD/SEM-EDX verify replacement of carbonate peaks by fluor- or carbonate-fluorapatite, with residual quartz acting as an inert gangue. Fe_2O_3 + Al_2O_3 levels fall below the 3 wt% threshold critical for both wet-process acidulation and electrothermic operation.
(b) In all leach liquors, >65 wt% of dissolved solids are calcium acetate, admixed with small fractions of magnesium and aluminum acetates. After polishing filtration, these streams conform to food-grade specifications for calcium and magnesium propionates/acetates, opening an ancillary revenue path as antimicrobials, de-icing agents, or feed-preservation additives.

5. Industrial and scientific implications

The organic-acid beneficiation strategy demonstrated here reconciles *selectivity* (via mild acids), *sustainability* (closed-loop acetate recovery), and *economics* (dual-product valorization). It widens the exploitable resource base by upgrading ores once deemed sub-economic, and offers flexible feedstock options for both yellow-phosphorus furnaces and wet-process phosphoric-acid lines.

Future work should address continuous counter-current leaching, solvent extraction of the acetate liquor to regenerate acetic acid, and techno-economic optimization at pilot scale. Incorporating the suggested comparative process-economics chapter—complete with discounted-cash-flow analysis, life-cycle assessment, and benchmark scenarios—will further enhance the monograph's utility to practitioners and investors in the phosphorus industry.

References

1. Government of the Republic of Kazakhstan (2012) Concept for the development of the geology sector of the Republic of Kazakhstan until 2030. Approved by Resolution No. 1042 dated August 13, 2012. https://adilet.zan.kz/kaz/docs/P1200001042. Accessed 3 Jan 2025
2. Dar SA, Khan KF, Birch WD (2017) Sedimentary: phosphates. In: Reference module in earth systems and environmental sciences. Elsevier, Amsterdam, 17 p
3. Mikhailova IA, Bondarenko OB (2006) Paleontology, 2nd edn. Moscow State University Press, Moscow, 592 p
4. Earle S (2015) Physical geology. Vancouver Island University, Nanaimo, 336 p
5. Van Kauwenbergh SJ (2010) World phosphate rock reserves and resources. IFDC, Alabama, 60 p
6. He G, Zhou Y (2015) Geology of phosphate rock in China: distribution, rock type and metallogenic perspective. Glob Environ Res 19:91–96
7. Filko AS, Faizulin RM (1994) Phosphate ores of Russia. Miner Resour Russia 5:18–25
8. US Geological Survey. https://www.usgs.gov/media/images. Accessed 3 Feb 2025
9. Daafi Y, Chakir A, Jourani E et al (2014) Geology and mine planning of phosphate deposits: Benguerir deposit Gantour Basin—Morocco. Procedia Eng 83:70–75
10. Jinning phosphate deposit. Jinning Co., Kunming. https://www.mindat.org/loc-155757.html. Accessed 7 Mar 2025
11. Phosphate. https://energymining.sa.gov.au/minerals/mineral_. Accessed 7 Mar 2025
12. State report on the condition and use of mineral resources of the Russian Federation in 2012. https://www.ecoindustry.ru/gosdoklad/view/263.html?ysclid. Accessed 7 Mar 2025
13. Al-Hobaib A, Baioumy HM, Al-Ateeq MA (2013) Geochemistry and origin of the Paleocene phosphorites from the Hazm Al-Jalamid area, northern Saudi Arabia. J Geochem Explor 132:15–25
14. Shibutov M (2017) Strategic deposits of Kazakhstan: evaluation and list. https://regnum.ru/news/economy/2262120.html. Accessed 11 Mar 2025
15. Huge phosphate discovered in Guizhou province. http://russian.people.com.cn/n3/2017/0713/c31518-9241358.html. Accessed 11 Mar 2025
16. Elkondieva GB (2012) Karatau basin—the largest phosphate raw material base of Eurasia. In: Proceedings of the international scientific and practical conference on phosphate raw materials: production and processing, Moscow, pp 21–30
17. Kiperman YA (2006) Phosphates in the 21st century: monograph. Almaty; Taraz; Zhanatas, 208 p
18. Bushinsky GI (1966) Ancient phosphorites of Asia and their genesis. Nauka, Moscow, 230 p

Y. Raiymbekov, *Enrichment of Low-Grade Phosphorites with Organic Acids*,
https://doi.org/10.1007/978-3-032-02910-2

19. Solid mineral resources. http://info.geology.gov.kz/ru/informatsiya/spravochnik-mestorozhdenij-kazakhstana/tverdye. Accessed 13 Mar 2025
20. Gimmelfarb BM (1965) Patterns of distribution of phosphate deposits of the USSR and their genetic classification. Nedra, Moscow, 308 p
21. Smirnov AI (1974) Material composition of phosphate ores of major industrial deposits of the USSR. Geol Phosphate Deposits 26:84–101
22. Hakkou R, Benzaazoue M, Bussiere B (2016) Valorization of phosphate waste rocks and sludge from the Moroccan phosphate mines: challenges and perspectives. Procedia Eng 138:110–118
23. Karatau mining and processing complex, Kazphosphate LLP. http://www.kpp.kz/kaz/stru_podr/gpkk/index.php. Accessed 17 Mar 2025
24. Phosphate flour Chilisai—a highly effective, environmentally friendly slow-release phosphate mineral fertilizer. https://chilisai.kz/index.php?lang=ru. Accessed 17 Mar 2025
25. Raiymbekov EB, Besterekov U, Abdurazova PA et al (2020) Low-grade raw materials and waste of Karatau phosphorites as industrial commodity raw materials. Bull KazNTU 1(137):664–668
26. Besterekov U, Nazarbek U, Raiymbekov Y et al (2021) Resource indicators of phosphate raw materials in the world. In: Proceedings of the 7th international conference on industrial technologies and engineering (ICITE—2021). M. Auezov South Kazakhstan University, Shymkent, pp 60–62
27. Hartman HL, Mutmansky JM (2002) Introductory mining engineering. John Wiley & Sons, New Jersey, 592 p
28. Chang JS, Kelly A, Crowley JM (1995) Handbook of electrostatic processes. CRC Press, Boca Raton, 780 p
29. Swain AK, Patra H, Roy GK (2011) Mechanical operations, 1st edn. Tata McGraw-Hill Education, New York, 540 p
30. Fuerstenau MC, Han KN (2003) Principles of mineral processing. SME, Southfield, 573 p
31. Komarova ZA (2012) Material composition and beneficiability of a micrograined phosphate ore sample from the Koksu deposit (Kazakhstan). Min Inf Anal Bull 9:379–388
32. Kojevnikov AO, Struzhkov VI (1992) Method of flotation-chemical beneficiation of natural phosphorites. RU Patent 1780836. Published 15 Dec 1992, Bull No. 46, 6 p
33. Kojevnikov AO, Struzhkov VI (1992) Method of chemical-flotation beneficiation of natural phosphorites. RU Patent 1773491. Published 7 Nov 1992, Bull No. 41, 8 p
34. Baichenko AA, Klein MS (1982) Method for beneficiation of carbonate-siliceous phosphate ores. RU Patent 956022. Published 7 Sep 1982, Bull No. 33, 2 p
35. Albuquerque RO (2012) Flotation routes for a phosphate ore bearing silicate-carbonate gangue. Rev Fac Ing 27:26–32
36. Zheng X, Smith RW (2000) Dolomite depressants in the flotation of apatite and collophane from dolomite. Miner Eng 10(5):537–545
37. Abdel-Khalek NA (2000) Evaluation of flotation strategies for sedimentary phosphates with siliceous and carbonate gangue. Miner Eng 13(7):789–793
38. Boulos TR, Abdel-Khalek NA, Ibrahim SS (2000) Rationalization of the flotation circuit of an Egyptian phosphorite plant. Tenside Surfact Det 3:176–182
39. Boulos TR, Yehia A, Ibrahim SS, Yassin KE (2014) A modification in the flotation process of a calcareous–siliceous phosphorite that might improve the process economics. Miner Eng 69:97–101
40. Lawver JE, Raulerson JD, Cook CC (1980) New techniques in beneficiation phosphate rock. Trans SME-AIME 268:1787–1801
41. Houot R, Joussemet R, Tracez J, Brouard R (1985) Selective flotation of phosphatic ores having a siliceous and/or a carbonated gangue. Int J Miner Process 14:245–264
42. Cheng X, Li R, Ju C et al (2013) Study on collector for reverse flotation of certain phosphorite in Guizhou. Adv Mater Res 734–737:1086–1092
43. Zhang JS, Que XL (2008) Mining agents. Press of Metallurgy Industry, Beijing, 350 p

44. Pharmaceutical Group in Lab 2, Chemical Mining Design and Research Institute of Ministry of Chemical Industry (1975) The application of SDS in flotation experiments of Yichang phosphate rock. Ind Miner Process 1:36–38
45. Kun S, Tao L, Yimin Z et al (2017) Application and mechanism of anionic collector sodium dodecyl sulfate (SDS) in phosphate beneficiation. Minerals 7(2):29-1–29-13
46. Seth V, Kumar R, Arora SCD et al (1975) Disodium dodecyl phosphate as a collector in the calcite–apatite mineral system. Trans Inst Min Metall 84:56–58
47. Pinto CAF, Yarar B, Araujo AC (1991) Apatite flotation kinetics with conventional and new collectors. In: Proceedings of the SME annual meeting, Denver, Colorado, pp 80–91
48. Assis SM, Montenegro LCM, Peres AAC (1996) Utilization of hydroxamates in minerals froth flotation. Miner Eng 9(1):103–114
49. Buttner BM, Grote J, Vogt V et al (1991) Flotation and flocculation of ultrafine-grained apatite ores. In: Proceedings of the 17th international mining process congress, vol 1. Bergakademie, Freiberg, pp 71–82
50. Miller JD et al (2001) A new collector chemistry for phosphate flotation. In: Proceedings of the SME annual meeting, Denver, Colorado, pp 1–46
51. Clerici C, Morandini AF, Mancini A et al (1984) Flotation of phosphate rock with carbonate-quartz gangue. In: Reagents in the minerals industry. Institution of Mining and Metallurgy, London, pp 221–225
52. Houot R, Joussemet R, Tracez J, Brouard R (1985) Selective flotation of phosphate ores having a siliceous and/or a carbonated gangue. Int J Min Process 14:245–264
53. Shao X, Bao Y, Guo M (1993) Adsorption properties of dissolved ions and their effects on the separation of collophane and dolomite. In: Beneficiation of phosphate: theory and practice. Society for Mining, Metallurgy, and Exploration, Littleton, pp 267–284
54. Shao X, Jiang CL, Parekh PK (1998) Enhanced flotation separation of phosphate and dolomite using a new amphoteric collector. SME Min Metall Process 15(2):11–14
55. Rao KH, Su F, Forssberg KSE (1999) Flotation kinetics of apatite from magnetite. In: Beneficiation of phosphates: advances in research and practice. Society for Mining, Metallurgy, and Exploration, Littleton, pp 103–125
56. Matiolo E, Couto HJB, de Lira Teixeira MF et al (2019) A comparative study of different columns sizes for ultrafine apatite flotation. Minerals 9:391-1–391-11
57. Clifford P, Lloyd M, Zhang P (1998) Technology research improves phosphate economics. Min Eng 98:46–51
58. Baudet G (1988) The processing of phosphate ores. Chron Rch Miner, special issue on phosphates:67
59. Silvia AF, Andery PA (1972) Phosphorus and potassium. Br Sulphur Corp 57:37–40
60. Kunii D, Levenspiel O (1969) Fluidization engineering. John Wiley & Sons, Weinheim, 630 p
61. Alperovich IG (1982) Phosphate raw materials and their thermal preparation for yellow phosphorus sublimation: monograph. LenNIIGiprokhim, Leningrad, 147 p
62. Coffey RD, Mooney KW, Cromwell GI et al (1994) Biological availability of phosphorus in defluorinated phosphates with different phosphorus solubilities in neutral ammonium citrate for chicks and pigs. Anim Sci 72:2653–2660
63. Ben-Ari C, Fuchs WJ (1963) Beneficiation of natural phosphate containing calcite. UK Patent 908018. Published 12 Aug 1963, Bull No. 3, 12 p
64. Ben-Ari C, Fuchs WJ (1963) Beneficiation of natural phosphate containing calcite. UK Patent 908137. Published 13 Jun 1963
65. Herman ER (1966) Beneficiation of natural phosphates containing calcite and resulting products. FR Patent 1398563. Published 18 Dec 1966
66. Borisov VM, Fridman NG, Smirnov AI, Zalavina TP (1974) Method for beneficiation of phosphorite ore. SU Author's Certificate 390831. Published 14 Apr 1974
67. Flaszenberg A, Prulov I, Lapidot M (1964) Method for beneficiation of natural phosphate. US Patent 3067023. Published 18 Mar 1964

68. Jerzy W (1967) Method of processing low-percentage Polish phosphorites into phosphoric acid and concentrated phosphoric fertilizers. Przem Chem 7(46):369–373
69. Rambabu C, Majumdar KK, Phadnis AN (1973) Study of the enrichment of phosphate ore containing calcite from the Kanpur deposit (Rajasthan). Indian J Technol 11(2):78–82
70. Smirnov YM, Mash'yanova AV, Aizenshtat MD et al (1977) Improving the beneficiation efficiency of phosphate-carbonate ores by roasting. Trans Inst Min Chem Raw Mater 39:87–92
71. Priestley RJ (1978) Beneficiation of phosphate rock. US Patent 4083929. Published 11 Apr 1978, 4 p
72. Kumar D (1980) Calcination of phosphate ores. Chem Ing Techn 52(9):739–740
73. Gunduz T, Gumgum B (1987) Enrichment of low-quality phosphates of the Mazidagi deposit by calcination and extraction methods. Sep Sci Technol 22(6):1645–1648
74. Emich GD (1984) Phosphate rock. Ind Miner Rocks 2:1017–1047
75. Disanto BJ (1985) Pyrometallurgy. In: SME mineral processing handbook. Metallurgical and Petroleum Engineers, Inc., New York, 59 p
76. Henin JP, Lectard A (1983) Suspension flash calcining. Eng Min J 10:77–84
77. Henin JP, Pinoncely A (1986) FCB and flash calcination. Ind Miner Mines Carrières Technol 6:249–252
78. Mbissik A, Elghali A, Ouabid M et al (2021) Alkali-hydrothermal treatment of K-rich igneous rocks for their direct use as potassic fertilizers. Minerals 11:140-1–140-22
79. Abbes N, Bilal E, Hermann L et al (2020) Thermal beneficiation of Sra Ouertane (Tunisia) low-grade phosphate rock. Minerals 10:937-1–937-13
80. Khaddor M, Ziyad M, Joffre J et al (2002) Pyrolysis and characterization of the kerogen from the Moroccan Youssoufia rock phosphate. Chem Geol 186(1–2):17–30
81. Al-Fariss RF (1993) Beneficiation of carbonate-rich Saudi phosphate rocks. In: Beneficiation of phosphate: theory and practice. SME, Phoenix, pp 251–259
82. Elgilliani DA, Abouzeid AZM (1993) Flotation of carbonates from phosphate ores in acidic media. Int J Miner Process 38(3–4):235–256
83. White JC, Goff TN, Good PC (1978) Continuous-circuit preparation of phosphoric acid from Florida phosphate matrix. BuMines Rl, Washington, DC, 22 p
84. Issahary D (1982) Prediction of the phosphorus pentoxide content in the product as a function of raw material composition in phosphate beneficiation by calcination. Anal Chim Acta 138:183–190
85. Wu LLR, Shibin Y (1985) Concentration of dolomitic phosphate rock and recovery of iodine at Wengfu phosphorus mine. In: Proceedings of the 15th international mineral processing congress, Cannes, pp 400–411
86. Good PC (1976) Beneficiation of unweathered Indian calcareous phosphate rock by calcination and hydration. US Bureau of Mines, Washington, DC, 17 p
87. El-Jalead IS, Abouzeid AZM, El-Sinbawy HA (1980) Calcination of phosphates: reactivity of calcined phosphate. Powder Technol 26:187–197
88. Potashnik BA, Avakyan ZA, Karavaiko GI et al (1997) Method for beneficiation of carbonate-containing phosphate raw materials. RU Patent 2097139. Published 27 Nov 1997, 5 p
89. Costa ACA, Medronho RA, Pecanha RP (1992) Phosphate rock bioleaching. Biotechnol Lett 14:233–238
90. Kokal H (1990) The origin of phosphorus in ironmaking raw materials and methods of removal: a review. In: Proceedings of the 63rd annual meeting Minnesota section AIME, Englewood, pp 225–257
91. Rodriguez H, Fraga R (1999) Phosphate solubilizing bacteria and their role in plant growth promotion. Biotechnol Adv 17:319–339
92. Igual JM, Valverde A, Cervantes E et al (2001) Phosphate solubilizing bacteria as inoculants for agriculture: use of updated molecular techniques in their study. Agronomie 21:561–568
93. Pyagai AG, Beglov BM, Kim Lin Zu, Borukhov IA (1990) Method for beneficiation of phosphorites. SU Patent 1555315. Published 4 Jul 1990, Bull No. 13, 3 p

94. Tleukulov OM, Kim Lin Zu, Buketov EA (1970) Method for beneficiation of phosphorites. SU Patent 290025. Published 22 Dec 1970, Bull No. 2, 1 p
95. Tleukulov OM, Dzhurumbaev AI, Berzhanov DS et al (1990) Method for beneficiation of phosphorites. SU Patent 1585310. Published 15 Aug 1990, Bull No. 30, 3 p
96. Treushchenko NN, Lavrova TV, Belyakov VA et al (1983) Method for beneficiation of magnesium-containing phosphate raw materials. SU Patent 1058878. Published 7 Dec 1983, Bull No. 45, 5 p
97. Wilemon GM, Scheiner BJ (1979) Leaching of the phosphate values from two Central Florida ores, using H2SO4-methanol mixtures. Report 9094. Bureau of Mines, Washington, DC, 9 p
98. Dahanayake K, Senaratne A, Subasinghe S et al (1991) Potential use of naturally occurring sulfuric acid to beneficiate poorly soluble phosphate from Eppawala, Sri Lanka. Fertil Res 29:197–201
99. Good PC, Goff TN, White JC (1979) Acidulation of Florida phosphate matrix in a single-tank reactor. Report 8339. Bureau of Mines, Washington, DC, 16 p
100. Liu V (1989) Method for beneficiation of phosphate-containing ore. SU Patent 1516006. Published 15 Oct 1989, Bull No. 38, 9 p
101. Chaikina MV (1977) Method for beneficiation of carbonate-containing phosphorites. SU Patent 559915. Published 30 May 1977, Bull No. 20, 2 p
102. Sikharulidze NG, Kapanadze ES, Tskhakaya NS (1988) Method for beneficiation of phosphate-containing ores. SU Patent 1392063. Published 30 Apr 1988, Bull No. 16, 3 p
103. Grigoryan GO (1975) Method for beneficiation of phosphorites. SU Patent 469664. Published 5 May 1975, Bull No. 17, 2 p
104. Petukhov OF, Shtoller VV (1982) Method for beneficiation of magnesium-containing phosphorites. SU Patent 937329A1. Published 23 Jun 1982, Bull No. 23, 2 p
105. Zafar IZ, Anwar MM, Pritchard DW (2006) Selective leaching of calcareous phosphate rock in formic acid: optimisation of operating conditions. Miner Eng 19(14):1459–1461
106. Zafar IZ, Anwar MM, Pritchard DW (1996) A new route for the beneficiation of low-grade calcareous phosphate rocks. Fertil Res 44(2):133–142
107. Zafar IZ, Ashraf M (2007) Selective leaching kinetics of calcareous phosphate rock in lactic acid. Chem Eng J 131(1):41–48
108. Ashraf M, Zafar ZI, Ansari TM (2005) Selective leaching kinetics and upgrading of low-grade calcareous phosphate rock in succinic acid. Hydrometallurgy 80(4):286–292
109. Sadeddin W, Abu-Eishah SI (1990) Minimization of free calcium carbonate in hard and medium-hard phosphate rocks using dilute acetic acid solution. Int J Miner Process 30(1):113–125
110. Abu-Eishah SI, El-Jallad IS, Muthaker M et al (1991) Beneficiation of calcareous phosphate rocks using dilute acetic acid solutions: optimisation of operating conditions for Ruseifa (Jordan) phosphate. Int J Miner Process 31(1):115–126
111. Zafar IZ (1993) Beneficiation of low-grade carbonate-rich phosphate rocks using dilute acetic acid solution. Fertil Res 34(2):173–180
112. Sengul H, Ozer AK, Gulaboglu MS (2006) Beneficiation of Mardin-Mazidagi (Turkey) calcareous phosphate rock using dilute acetic acid solution. Chem Eng J 122(3):135–140
113. Bakry AR, Abdelfattah NA, Farag AB et al (2015) Upgrading of Abu-Tartur calcareous phosphate via selective leaching by organic acids. Int J Eng Res 6(10):57–64
114. Fei X, Jie Z, Jiyan C et al (2019) Research on enrichment of P2O5 from low-grade carbonaceous phosphate ore via organic acid solution. J Anal Methods Chem 2019:9859580-1–9859580-7
115. Haweel CK, Abdul-Majeed BA, Eisa MY (2013) Beneficiation of Iraqi Akash at phosphate ore using organic acids for the production of wet process phosphoric acid. Al-Khwarizmi Eng J 9(4):24–38
116. Arroug L, Elaatmani M, Zegzouti A et al (2021) Low-grade phosphate tailings beneficiation via organic acid leaching: process optimization and kinetic studies. Minerals 11:492-1–492-18
117. Economou ED, Vaimakis TC (1997) Beneficiation of Greek calcareous phosphate ore using acetic acid solutions. Ind Eng Chem Res 36(5):1491–1497

118. Economou ED, Vaimakis TC, Papamichael EM (1998) Kinetics of dissolution of the carbonate minerals of phosphate ores using dilute acetic acid solutions. J Colloid Interface Sci 245(1):164–171
119. Economou ED, Vaimakis TC, Papamichael EM (2002) The kinetics of dissolution of the carbonate minerals of phosphate ores using dilute acetic acid solutions: the case of pH range from 3.96 to 6.40. J Colloid Interface Sci 245(1):133–141
120. Seitnazarov A, Namazov S, Beglov B (2014) Beneficiation of high-calcareous phosphorites of central Kyzylkum with organic acid solutions. J Chem Technol Metall 49(4):383–390
121. Gharabaghi M, Irannajad M, Noaparast M (2010) A review of the beneficiation of calcareous phosphate ores using organic acid leaching. Hydrometallurgy 103(1):96–107
122. Abouzeid AZM (2008) Physical and thermal treatment of phosphate ores: an overview. Int J Miner Process 85(4):59–84
123. Fredd CN, Fogler HS (1998) The kinetics of calcite dissolution in acetic acid solutions. Chem Eng Sci 53:3863–3874
124. Jordan G, Higgins SR, Eggleston CM et al (2001) Dissolution kinetics of magnesite in acidic aqueous solution, a hydrothermal atomic force microscopy (HAFM) study: step orientation and kink dynamics. Geochim Cosmochim Acta 23:4257–4266
125. Zafar IZ, Saeed A (2004) Mass transfer and reaction kinetics in leaching of calcareous phosphate rock. J Eng Horizons 17(4):15–21
126. Oral L, Bunyamin D, Fatih D (2005) Dissolution kinetics of natural magnesite in acetic acid solutions. Int J Miner Process 75:91–99
127. Demir F, Donmez B, Colak S (2003) Leaching kinetics of magnesite in citric acid solutions. J Chem Eng Jpn 36:683–688
128. Raiymbekov YB, Besterekov U, Abdurazova PA (2020) Review of methods for enrichment of phosphate raw materials in the world. Bull Karaganda Univ Chem Ser 2(98):92–96
129. Raiymbekov Y, Besterekov U, Abdurazova P et al (2022) Review of methods and technologies for the enrichment of low-grade phosphorites. https://www.degruyter.com/document/doi/10.1515/revic-2021-0041. Accessed 31 Jan 2025
130. Borovikov VP (2003) Statistica. The art of data analysis on a computer, 2nd edn. PITER, St. Petersburg, 688 p
131. Korovkin MV, Ananyeva LG (2016) Infrared spectroscopy of carbonate rocks and minerals. TPU, Tomsk, 75 p
132. Tarasevich BN (2012) IR spectra of major classes of organic compounds: reference material. Lomonosov MSU, Moscow, 55 p
133. Nakamoto K (1991) Infrared and Raman spectra of inorganic and coordination compounds. John Wiley & Sons, New York, 504 p
134. Papko LF, Kravchuk AP (2013) Physicochemical methods for studying inorganic substances and materials. BSTU, Minsk, 100 p
135. Bukovshin VV (1999) Modern methods of studying mineral matter: thermal and thermoluminescence analysis. VSU, Voronezh, 38 p
136. Ptacek P (2016) Apatites and their synthetic analogues: synthesis, structure, properties and applications. IntechOpen, London, p 512
137. Nriagu JO, Moore PB (1984) Phosphate minerals. Springer-Verlag, New York, 450 p
138. Yoder CH (2006) Ionic compounds. Applications of chemistry to mineralogy. John Wiley & Sons, New Jersey, 207 p
139. Petruk W (2000) Applied mineralogy in the mining industry. Elsevier, Amsterdam, 287 p
140. Bulakh AG, Zolotarev AA, Krivovichev VG (2008) General mineralogy: textbook. Akademiya, Moscow, 416 p
141. KZ ST 2211-2012. Karatau fine-ground phosphate raw material. Technical specifications
142. KZ ST 2213-2012. Karatau powdered phosphate raw material. Technical specifications
143. Dorfman MD, Bussen IV, Dudkin OB (1959) Some data on selective dissolution of minerals. Trans Miner Museum 9:167–171

144. Vargas C (2017) Organic acids: characteristics, properties and synthesis. Nova Science Publishers, New York, 144 p
145. Raiymbekov Y, Besterekov U, Abdurazova P et al (2021) Beneficiation of phosphate-siliceous slates via acetic acid. Int J Chem React Eng 19(11):1187–1195
146. Besterekov U, Raiymbekov YB, Nazarbek UB et al (2021) Method for beneficiation of off-grade phosphorites. KZ Patent 6588. Published 29 Oct 2021, Bull No. 43, 4 p
147. Beletsky VS (2004) Small mining encyclopedia. Donbass, Donetsk, 640 p
148. Gharabaghi M, Noaparast M, Irannajad M (2009) Selective leaching kinetics of low-grade calcareous phosphate ore in acetic acid. Hydrometallurgy 95:341–345
149. Taghavi M, Gharabaghi M, Shafaei SZ (2020) Selective leaching of low-grade phosphate ore using a mixture of organic acids. Int J Min Geo-Eng 54:65–70
150. Voldman GM, Zelikman AN (2003) Theory of hydrometallurgical processes: textbook, 4th edn. Intermet Engineering, Moscow, 464 p
151. Ryzhenkov DI, Arsentyev PP, Yakovlev VV (1989) Theory of metallurgical processes. Metallurgy, Moscow, 352 p
152. Heidarpour T (2009) Processing of Dalir phosphate samples using leaching method. MSc thesis, Amirkabir University of Technology
153. Jander W (1927) Reaktionen im festen Zustande bei höheren Temperaturen. Reaktionsgeschwindigkeiten endotherm verlaufender Umsetzungen. Z Anorg Allg Chem 163:1–30
154. Makanyire T, Jha A, Sutcliffe S (2016) Kinetics of hydrochloric acid leaching of niobium from TiO_2 residues. Int J Miner Process 157:1–6
155. Raiymbekov Y, Abdurazova P, Nazarbek U (2023) Enrichment of low-grade phosphorites by the selective leaching method. Green Process Synth 12:20228150
156. Raiymbekov YB, Abdurazova PA, Nazarbek UB (2023) Thermodynamic analysis of enrichment of low-grade phosphate raw materials with organic acid. Gorn Zhurnal Kazakhstana 5:32–38
157. Raiymbekov Y, Abdurazova P, Nazarbek U (2022) The main deposits of phosphorites in the Republic of Kazakhstan. In: Proceedings of the IX international conference on industrial technologies and engineering (ICITE—2022), vol I. South Kazakhstan University, Shymkent, pp 94–97
158. Raiymbekov YB, Abdurazova PA, Nazarbekova SP, Nazarbek UB (2024) Enhancing the concentration of nodular phosphorite by selective leaching. News Natl Acad Sci Kazakhstan Chem Technol Ser 1(458):116–136
159. Raiymbekov EB, Abdurazova PA, Nazarbek UB, Nazarbekova SP (2024) Method for enrichment of low-grade phosphorites with citric acid. KZ Utility Model Patent No. 9386. Published 19 Jul 2024
160. Raiymbekov Y, Abdurazova P, Nazarbek U (2024) Dissolution kinetics of carbonates of low-grade microgranular phosphate ore using organic acids as leaching agent. Mining 4(4):766–776
161. Evenchik SD, Brodsky AA (1987) Technology of phosphate and complex fertilizers. Khimiya, Moscow, 464 p
162. Ershov VA, Belov VN (1979) Phosphorus technology. Khimiya, Leningrad, 336 p
163. Mohamed MAK, Ibrahim G, Rizk A et al (2018) Economics of exploitation phosphate ore wastes. Int J Min Eng Miner Process 7(1):14–20
164. Carlson JT (2013) Beneficiation of high-MgO sedimentary phosphate ores. PhD thesis, Michigan Technological University, Houghton, 217 p

Index

Y. Raiymbekov, *Enrichment of Low-Grade Phosphorites with Organic Acids*, https://doi.org/10.1007/978-3-032-02910-2

MIX
Papier aus verantwortungsvollen Quellen
Paper from responsible sources
FSC® C105338

If you have any concerns about our products,
you can contact us on
ProductSafety@springernature.com

In case Publisher is established outside the EU,
the EU authorized representative is:
Springer Nature Customer Service Center GmbH
Europaplatz 3, 69115 Heidelberg, Germany

Printed by Libri Plureos GmbH
in Hamburg, Germany